CW01203381

1 MONTH OF
FREE
READING

at

www.ForgottenBooks.com

By purchasing this book you are eligible for one month membership to ForgottenBooks.com, giving you unlimited access to our entire collection of over 1,000,000 titles via our web site and mobile apps.

To claim your free month visit:

www.forgottenbooks.com/free315344

* Offer is valid for 45 days from date of purchase. Terms and conditions apply.

ISBN 978-0-365-21474-8
PIBN 10315344

This book is a reproduction of an important historical work. Forgotten Books uses state-of-the-art technology to digitally reconstruct the work, preserving the original format whilst repairing imperfections present in the aged copy. In rare cases, an imperfection in the original, such as a blemish or missing page, may be replicated in our edition. We do, however, repair the vast majority of imperfections successfully; any imperfections that remain are intentionally left to preserve the state of such historical works.

Forgotten Books is a registered trademark of FB &c Ltd.
Copyright © 2018 FB &c Ltd.
FB &c Ltd, Dalton House, 60 Windsor Avenue, London, SW19 2RR.
Company number 08720141. Registered in England and Wales.

For support please visit www.forgottenbooks.com

Vacuum Cleaning Systems

A Treatise on the Principles and Practice of Mechanical Cleaning

BY

M. S. COOLEY, M. E.

MECHANICAL ENGINEER IN OFFICE OF THE SUPERVISING ARCHITECT, TREASURY DEPARTMENT, WASHINGTON, D. C.

FIRST EDITION

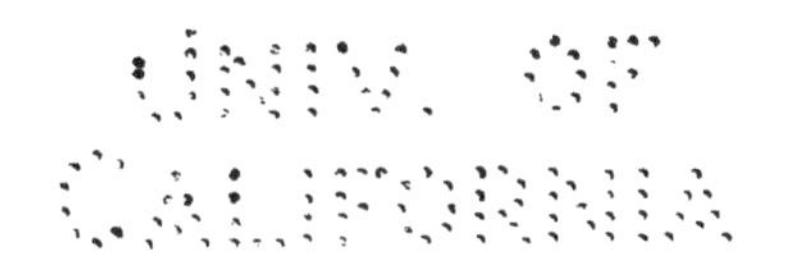

New York:
HEATING AND VENTILATING MAGAZINE COMPANY,
1123 Broadway

Copyright, 1913,

BY

HEATING AND VENTILATING MAGAZINE CO.

CHAPTER I.

HISTORY OF MECHANICAL CLEANING.

CHAPTER II.

REQUIREMENTS OF AN IDEAL VACUUM CLEANING SYSTEM.

CHAPTER III.

THE CARPET RENOVATOR.

302929

CHAPTER IV.

OTHER RENOVATORS.

CHAPTER V.

STEMS AND HANDLES.

CHAPTER VI.

HOSE.

CONTENTS

CHAPTER VII.

Pipe and Fittings.

CHAPTER VIII.

Separators.

CHAPTER IX.

Vacuum Producers.

CHAPTER X.

CONTROL.

CHAPTER XI.

SCRUBBING SYSTEMS.

CHAPTER XII.

SELECTION OF CLEANING PLANT.

CHAPTER XIII.

Tests.

CHAPTER XIV.

Specifications.

CHAPTER XV.

Portable Vacuum Cleaners.

TABLES.

ILLUSTRATIONS.

PREFACE.

The contents of this work are compiled from the observations of the author through the seven years during which he has been engaged in the preparation of specifications for, and the testing of, complete plants installed in the buildings under the control of the Treasury Department.

During this time it has become necessary to alter no less than five times the stock form of specifications for stationary vacuum cleaning plants which were adopted by the Government, with the intent of obtaining the widest competition possible with efficient and economical operation, in order to keep pace with the variation and improvement in the apparatus manufactured. As each new type of system has come on the market a personal investigation at the factory, together with tests, has been made. An exhaustive test of carpet renovators was also conducted, using one of the Government plants. In addition the vacometers recommended for use in capacity tests were carefully calibrated, using the machine at the Department of Agriculture.

The writer wishes to acknowledge the aid received from the various manufacturers in furnishing illustrations and data on their machines, to Messrs. Ewing & Ewing and Prof. Sidney A. Reeve for data on tests made by Prof. Reeve and used in defending the Kenney basic patent.

In analyzing the results of his tests and observations, the writer has endeavored to put his own conclusions into concrete form for the use of the consulting engineer and has not entered into the problems to be encountered in the design and manufacture of the various forms of apparatus.

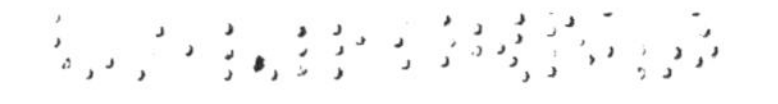

CHAPTER I.

HISTORY OF MECHANICAL CLEANING.

Early Attempts.—Whenever machinery has been introduced to assist or replace manual labor, the earlier attempts have been in imitating the tools formerly used by man. As the earliest mechanically-propelled carriages were mechanical walking machines, the earliest steamboats mechanical rowing machines, and the earliest flying machines mechanical birds, so were the earliest mechanical cleaners in the form of mechanical brooms.

These mechanical brooms were introduced about 1880 and took the form of the well-known street sweeper, with a large circular brush mounted on a four-wheeled cart and rotated by means of gearing driven from the wheels, the propelling power being the horses which drew the machine.

This machine at once made itself unpopular with the residents of the streets cleaned on account of its great activity in stirring up dust, because the streets were swept dry. This trouble was later overcome to a considerable extent by sprinkling the streets before sweeping, but only at a sacrifice in efficiency of cleaning, especially where such uneven surfaces as cobble or medina stone blocks formed the surface of the roadway. Various attachments were added to reduce this dust nuisance, but none has apparently been successful, as we see these machines in their original form in use today.

Almost simultaneously with the introduction of the street sweeper came its counterpart, the carpet sweeper, with a similar but smaller brush, enclosed in a wood and metal case, the brush being driven by friction from the wheels supporting the box and the power for operation being derived from the person who pushed the machine along the floor.

This machine has not been modified to any great extent during the thirty odd years of its existence. It is today in praetically its original form, and is doing no better work than when first introduced. This form of mechanical cleaner occupied the field of household cleaning for nearly twenty years without a rival, during which time it won its way into the hearts and hands of many housekeepers in this and other countries.

Limitations of the Carpet Sweeper.—This device, with its light brush and equally light pressure on the surface cleaned and its limited capacity for carrying the material picked up, has never been a thorough cleaner in any sense of the word, and has been and is now used only to take up that portion of the usual litter and light dust which is located directly on the surface, and is, therefore, most annoying to the housekeeper, owing to its being visible to the eye. Because of its generous proportions, made necessary to accommodate the material picked up, and its centrally-pivoted handle, made necessary by its mechanical construction, it is impossible to operate it under low furniture. Like the lawn mower, it must be in motion in order to operate its revolving brush, on which its cleaning action is dependent. It is impossible to make use of same in corners, along walls, or close to heavy furniture, its use being limited to a literal slicking up of those portions of the carpet in the most conspicuous portions of the apartment. In spite of these serious defects it came into, and is still in, nearly universal use, even in households equipped with the latest approved types of mechanical cleaners. Its use on bare floors has never been even a moderate success and in no case has it superseded the broom and dust pan of our grandmothers

Compressed Air Cleaners.— Compressed air has been in use for many years in foundries and machine shops, for cleaning castings and producing certain finishes on metal. With the introduction of modern electrical machinery it was rapidly adapted to the cleaning of windings and other inaccessible parts of this machinery. Its first use in cleaning buildings was undoubtedly in the form of an open jet for dislodging dust from carvings and relief work, for which purpose it is very efficient as a remover of the dust from the parts to be cleaned and also as a distributor of this same dust over the widest possible area for subsequent removal by other means. It has a draw-back in that the expansion of air both cools the same and reduces its ability to retain moisture, resulting in the deposit of moisture on the surfaces cleaned.

About 1898, attempts to overcome the objections to the open air jet and to produce a commercially successful compressed air carpet cleaner were undertaken almost simultaneously by two companies, the American Air Cleaning Company, of Milwaukee, operating under the Christensen patents, and the General Compressed Air Cleaning Company, of St. Louis, operating under the Thurman patents.

The renovator used by the American Air Cleaning Company consisted of a heavy metal frame, about 18 in. long and 12 in. wide, having mounted on its longer axis a wedge-like nozzle extending the entire length of the frame, with a very narrow slit, 1/64 in. wide, extending the entire length of its lower edge. This nozzle was pivoted and so connected to the operating handle, by which the renovator was moved over the floor, that when the renovator was alternately pushed and pulled over the surface to be cleaned, the slot was always inclined in the direction in which the renovator was being moved. The top of the renovator was closed by a canvas bag, smaller at the neck than in its center, which was supported by a wire hook.

Air was introduced into the nozzle, at a pressure of from 45 to 55 lbs. per square inch, and issued from the slot in a thin sheet which impinged on the carpet at an angle. The frame was held close to the carpet by its weight, preventing the escape of the air under its lower edge. The air striking the carpet at an angle was deflected up into the bag, inflating same like a miniature balloon. The dust loosened from the carpet by the impact of the air was carried up into the bag where it lodged, the air escaping through the fabric of the canvas into the apartment.

The renovator used by the General Compressed Air Cleaning Company differed from the above-described renovator in that it contained two nozzles, with slots inclined at fixed angles to the carpet. A pair of hand-operated valves were provided in the handle to introduce air into the nozzle which was inclined in the direction in which the renovator was moving; otherwise the renovator was identical with that used by the Milwaukee company.

These renovators were generally supplied with air from a portable unit, consisting of an air compressor, driven by a gasoline engine mounted with the necessary gasoline and air storage tanks on a small truck. One of these machines was in use in Washington last year, but its use at that time was very limited and it is not to be seen this year.

These trucks were drawn up in front of the building to be cleaned and a large-size hose, usually 1¼ in. in diameter, was carried into the house and attached to an auxiliary tank from which ½-in. diameter hose lines were carried to two or more renovators.

A few buildings were equipped with air compressors and pipe lines, with outlets throughout the building for use with this type of renovator, among which was the Hotel Astor in New York City.

These renovators, the construction of which is shown diagrammatically in Fig. 1, required approximately 35 cu. ft. of free

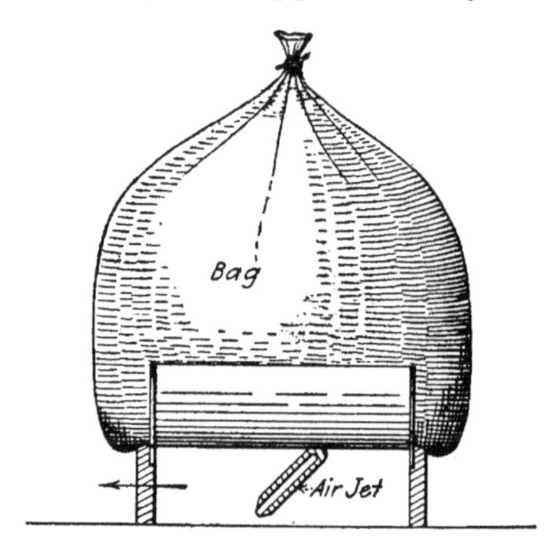

FIG. I. EARLY TYPE OF MECHANICAL CLEANING NOZZLE USING COMPRESSED AIR.

air per minute at a pressure of from 45 to 55 lbs. per square inch and were usually driven by a 15 H. P. engine.

The renovators were very heavy to carry about, although their operation with the air pressure under them was not difficult. However, their operation was complicated, requiring skilled operators. Owing to their generous proportions it was impossible to clean around furniture, making its removal from the apartment necessary, and limiting their use to the cleaning of carpets at the time of general house cleaning. The cooling effect of the expansion of the air in the nozzle often caused condensation of moisture on the carpets when the relative humidity was high. They were also at a disadvantage in that all the heavy dust collected in the canvas bag had to be carried from the apartment by hand. Owing to the constant agitation of the dust in the bag by the entering air currents, much of the finer particles of dust and all the disease germs liberated by the renovator were blown through the bag back into the apartment. They were not, therefore, by any means sanitary devices.

Vacuum Produced by Compressed Air.—The General Compressed Air Cleaning Company also introduced another form of renovator for use with their compressed air plants. This was composed of an ejector operated by compressed air, with a short hose attached to a carpet renovator of the straight narrow-slot type, such as was used later in vacuum cleaning systems. The outlet from this ejector was connected by another short hose to a metal box containing a canvas bag, woven backwards and forwards over metal frames to give a large surface for the passage of air. The dust picked up by the suction of the ejector was carried with the air into the box and there separated from the air, which escaped through the canvas into the apartment.

This form of renovator overcame some of the objections to the former type in that there was no condensation of moisture on the carpets, and it was possible to operate the renovator under and around furniture, and even on portieres and other hangings. However, the apparatus was rendered inefficient by the resistance of the bag, causing a back pressure on the injector which greatly reduced its air-drawing capacity.

Compressed Air Supplemented by Vacuum.—Shortly after these two companies began operation, the Sanitary Devices Manufacturing Company, of San Francisco, introduced a new system of mechanical cleaning under the Lotz patents. This system used a renovator having a compressed air nozzle terminating in a narrow slot, similar to the nozzles of the American and Thurman systems, but differing from them in that the slot was fixed vertically, pointing downward. This nozzle was surrounded by an annular chamber having an opening at the bottom of considerable width. The whole formed a renovator about 14 in. long and not over 2 in. wide at its base. In addition to the compressed air connection to its nozzle, a second hose, 1 in. in diameter, was connected to the annular space surrounding the nozzle and led to a vacuum pump by which the air liberated through the nozzle, together with the dust which was liberated from the carpet, was carried from the apartment. The construction of this renovator is shown diagrammatically in Fig. 2.

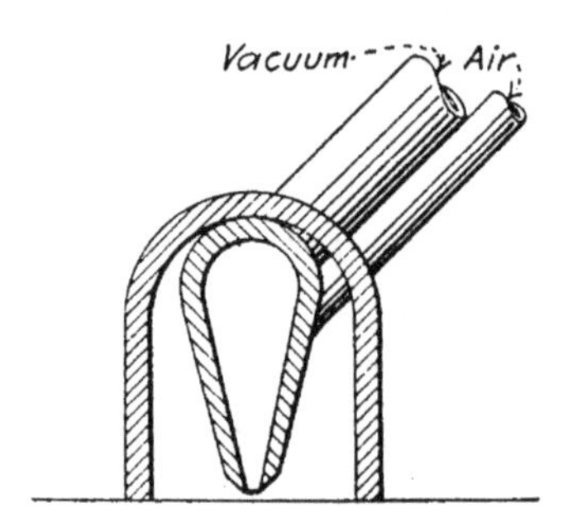

FIG. 2. ANOTHER TYPE OF COMPRESSED AIR CLEANING NOZZLE, SUPPLEMENTED WITH VACUUM PIPE.

As dust-laden air was not suitable to be carried through the pump used as a vacuum producer, separators had to be provided to remove the dust from this air before it reached the pump. The separators used consisted of two cylindrical tanks. The air was introduced into the first tank in such a way that a whirling motion was imparted to it, thus separating the heavier particles of dust by centrifugal force. The second tank contained water which was brought into intimate contact with the air by means of an atomizer located in the pipe connection between the two tanks, thus washing the air in a manner somewhat similar to the familiar air washers used in connection with mechanical ventilating systems. The air and spray then entered the second tank, above the water line, where the entrained water separated on the reduction of velocity and fell back into the water below, to be recirculated through the

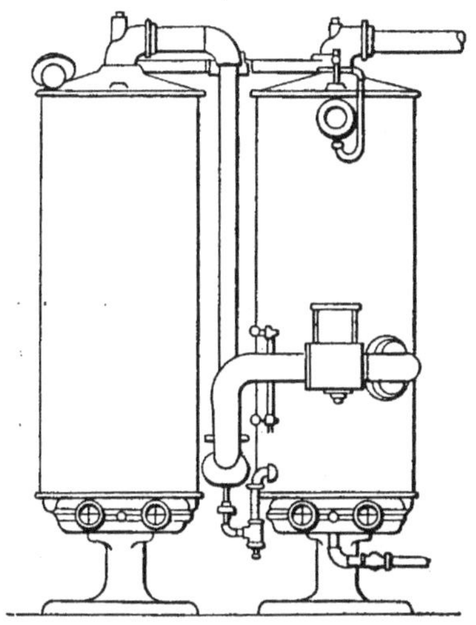

FIG. 3. SEPARATORS USED WITH COMBINED COMPRESSED AIR AND VACUUM MACHINES.

atomizer. The air passed on out of the top of the tank to the pump. An illustration of these separators is shown in Fig. 3

Piston Pump the First Satisfactory Vacuum Producer.—Various types of apparatus were tried as vacuum producers, including an air ejector, such as was used with the Thurman

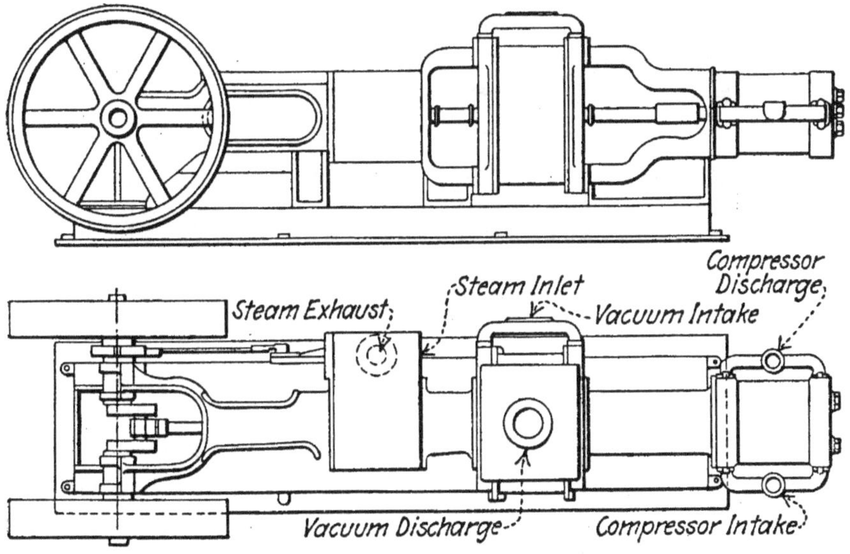

FIG. 4. PISTON TYPE OF VACUUM PUMP, MOUNTED TANDEM WITH AIR COMPRESSOR.

renovator, and found to be ineffective due to its inability to overcome the back-pressure necessary to discharge the air through the hose, which was placed on its outlet. A rotary pump was next tried, but, owing to the selection of an inefficient type, this was abandoned and, finally, a piston-type vacuum pump, with very light poppet valves and mounted tandem with the air compressor, was adapted and remained in use with this system until straight vacuum was adopted, when the air compression cylinder was omitted. This pump is illustrated in Fig. 4.

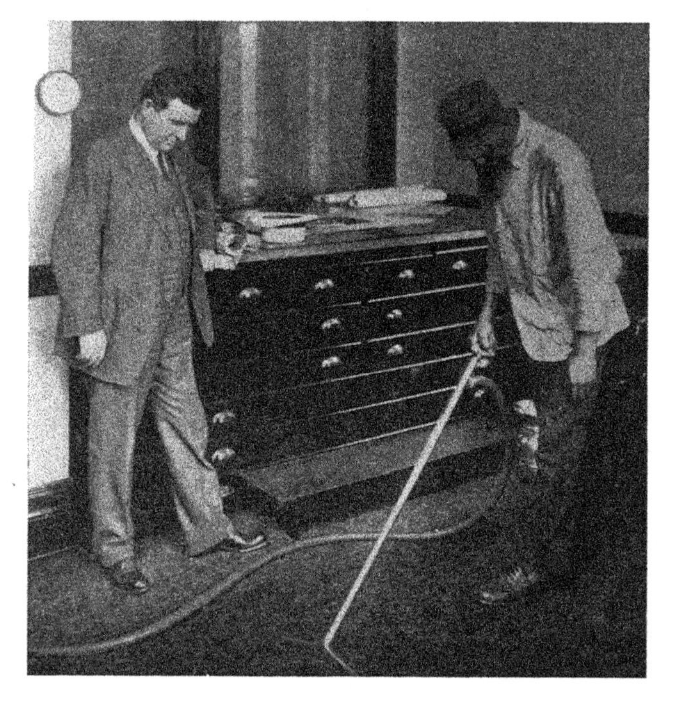

FIG. 5. MR. KENNEY'S FIRST RENOVATOR, VACUUM ALONE BEING USED AS CLEANING AGENT.

In this system we see the first sanitary device to be introduced into the field of mechanical cleaning, as the dust and germ-laden air were removed entirely from the apartment and purified before being discharged into the outside atmosphere. The foulness of the water in the separators clearly showed the amount of impurities removed from the air.

These machines were mounted on wagons, similar to their forerunners, and were also installed in many buildings as stationary plants, among which were the old Palace Hotel and the branch Mint, in San Francisco, and the old Fifth Avenue Hotel, in New York City.

Systems Using Vacuum Only.—In 1902 David T. Kenney, of New York, installed the first mechanical cleaning system in which vacuum alone was used as the cleaning agent. Mr. Kenney used a renovator with a slot about 12 in. long and 3/16 in. wide, attached to a metal tube which served as a handle,

FIG. 6. AIR COMPRESSORS ARRANGED FOR OPERATION AS VACUUM PUMPS.

and to a ¾-in. diameter hose and larger pipe line leading to separators and vacuum pump. Mr. Kenney's first renovator is illustrated in Fig. 5.

Mr. Kenney used as vacuum pumps commercial air compressors, the first of which was installed in the Frick Building in 1902 and is illustrated in Fig. 6. Later he adapted the Clayton air compressor, with mechanically-operated induction and poppet eduction valves on larger sizes, and single mechanically-operated induction and eduction valves on the smaller sizes.

The separators used by Mr. Kenney differed from those used by the Sanitary Devices Manufacturing Company in that they contained several interior partitions, screens, and baffles, and

FIG. 7. SEPARATORS INSTALLED BY MR. KENNEY IN FRICK BUILDING.

the air was drawn directly through the body of water in the wet separator. The relative merits of these types of separators will be discussed in a later chapter.

The separators installed by Mr. Kenney in the Frick Building, and which are practically the same as were used by him as long as he manufactured vacuum cleaning apparatus, are illustrated in Fig. 7.

After his application had been in the patent office for about six years he was granted a fundamental patent on a vacuum cleaning system.

Renovator with Inrush Slot.—The Sanitary Devices Manufacturing Company then produced a carpet renovator using vacuum only as a cleaning agent. This cleaner has a wider cleaning slot that the cleaners usually furnished by Mr. Kenney, about 5/16 in. wide, with a supplemental slot or vacuum breaker opening out of the top of the renovator and separated from the cleaning slot by a narrow partition extending nearly to the

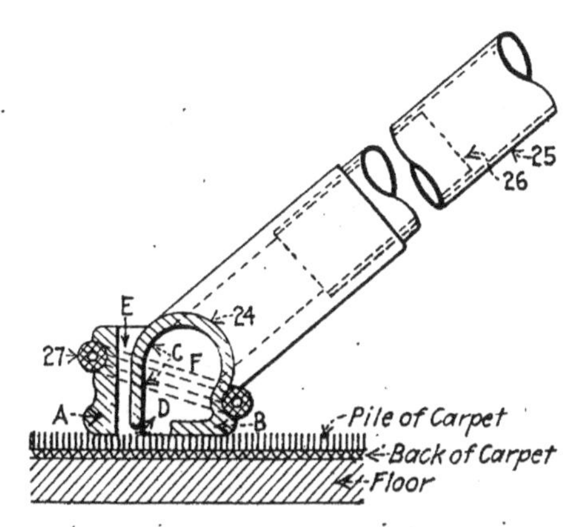

FIG. 8. VACUUM RENOVATOR WITH INRUSH SLOT, INTRODUCED BY THE SANITARY DEVICES MANUFACTURING CO.

carpet, as illustrated in Fig. 8. The relative merits of these types of renovators will be discussed in a later chapter.

Shortly after the introduction of vacuum cleaning by Mr. Kenney and the Sanitary Devices Manufacturing Company, the American Air Cleaning Company published an interesting little booklet entitled, "Compressed Air Versus Vacuum," which set forth in great detail the so-called advantages of compressed air over vacuum as a medium of mechanical carpet cleaning, and, apparently, proved that vacuum cleaners were much less efficient than cleaners operated by compressed air. A year or two later the American Air Cleaning Company evidently had a change of heart and began to manufacture these same "inefficient'' vacuum cleaners. Their previous treatise on vacuum cleaning, which apparently was not copyrighted, was republished by both the Sanitary Devices Manufacturing Company and by the Vacuum Cleaner Company, which had acquired Mr. Kenney's patents, and freely distributed. Thus this little work of the Milwaukee company, instead of injuring their competitors, was turned into good advertising for them and required a lot of explanation from the Milwaukee company.

Steam Aspirators Used as Vacuum Producers.—The American Air Cleaning Company used a steam aspirator as its vacuum producer and, unlike its predecessor, the air-operated ejector, it made good and has also been used to a limited extent by the Sanitary Devices Manufacturing Company. It is now marketed by the Richmond Radiator Company, and its merits will be discussed in a later chapter. The American Air Cleaner Company also used as a vacuum producer the single-impeller type of rotary pump, made by the Garden City Engineering Company, which was also later adopted, to a limited extent, by the Vacuum Cleaner Company. This will be discussed further on.

The renovator used by this company was a single-slot type, with ⅛-in. by 10-in. cleaning slot. These systems at once became notable on account of the small size of the vacuum producers used, the low degree of vacuum carried, and the vigorous campaign of advertising which was conducted.

Several firms soon began to market vacuum cleaning systems almost identical with that of Mr. Kenney, among which were the Blaisdell Machinery Company, The Baldwin Engineering Company, and The General Compressed Air and Vacuum Machinery Company, the latter being the original Thurman company.

The Vacuum Cleaner Company then began a series of infringment suits against nearly every manufacturer of vacuum cleaning systems. In nearly every case the suit has resulted in the offending company paying license fees to the Vacuum Cleaner Company, and this concern has now abandoned the manufacture of vacuum cleaners and has become a licensing company. At this writing nearly twenty firms are paying license fees to the Vacuum Cleaner Company and there is one suit now in the courts.

Piston Pump Used Without Separators.—A vacuum cleaning system of somewhat different design was produced by two former employees of the Vacuum Cleaner Company, Mr. Dunn, the once well-known "Farmer Dunn" of the weather bureau, afterward salesman for the Vacuum Cleaner Company, and Mr. Locke, at one time this firm's engineer. This company was first known as the Vacuum Cleaning Company, and, shortly afterward, as the Dunn-Locke Vacuum Cleaning Company. No separators were used with this system, but the dust-laden air was led from the pipe lines directly into a chamber on the pump, known as the "saturation chamber," and there mingled with a stream of water converting the dust into a thin mud. The air, water and mud then passed through the pump, the muddy water was discharged into the sewer, and the air into the atmosphere. The vacuum producer used was a piston pump without suction valves. With this system it was possible to handle water in almost unlimited quantities and with this feature a system of mechanical scrubbing was attempted for which great claims were made, none of which, however, were realized in a commercial way.

These gentlemen sold their patents to the E. H. Wheeler Company, which attempted to market the system in its original form. It was found, however, that the piston pump was not adapted to the handling of grit which was picked up by the renovators, and a rotary pump, with single impeller and a follower was substituted. This system is now marketed by the Vacuum Engineering Company, of New York, and is known as the Rotrex system.

Mr. Dunn again entered the field of vacuum cleaning and began marketing his machine a short time ago with a new form of automatic separator discharging to sewer.

First Portable Vacuum Cleaner.—About 1905, Dr. William Noe, of San Francisco, constructed the first portable vacuum cleaner. This machine contained a mechanically-driven rotary brush, similar to the brushes used in the familiar carpet sweeper, for loosening the dust from the carpet. This dust was sucked up by a two-stage turbine fan and discharged into a dust bag, mounted on the handle, similar to the bags on the compressed air cleaners. The whole machine was mounted on wheels and provided with a small direct-connected motor. This machine is illustrated in Fig. 9 and is the original form of the well-known Invincible renovator manufactured by the Electric Renovator Company, of Pittsburgh. This company now produces a complete line of stationary and portable vacuum cleaners, all of which use multi-stage turbines. The sale of the product of this company, until recently, was controlled by the United States Radiator Corporation.

FIG. 9. FIRST PORTABLE VACUUM CLEANER, CONSTRUCTED BY DR. WILLIAM NOE, OF SAN FRANCISCO, IN 1905.

First Use of Stationary Multi-Stage Turbine Blowers.— About 1905 Mr. Ira Spencer, president and engineer of the Organ Power Company, which manufactured a multi-stage turbine blower for organs, known as the "Orgoblow," organized the Spencer Turbine Cleaner Company and marketed a vacuum cleaning system, using a modification of the "Orgoblow" as a vacuum producer. These machines were first constructed with sheet metal casings and had sheet steel fans, with wings riveted on and mounted on horizontal shafts. The separators were sheet metal receptacles with screens for catching litter. Light-weight hose, 2 in. in diameter, was used to connect the renovators to 4-in. sheet metal pipe lines. A variety of renovators was produced for use with this system. Carpet renovators having

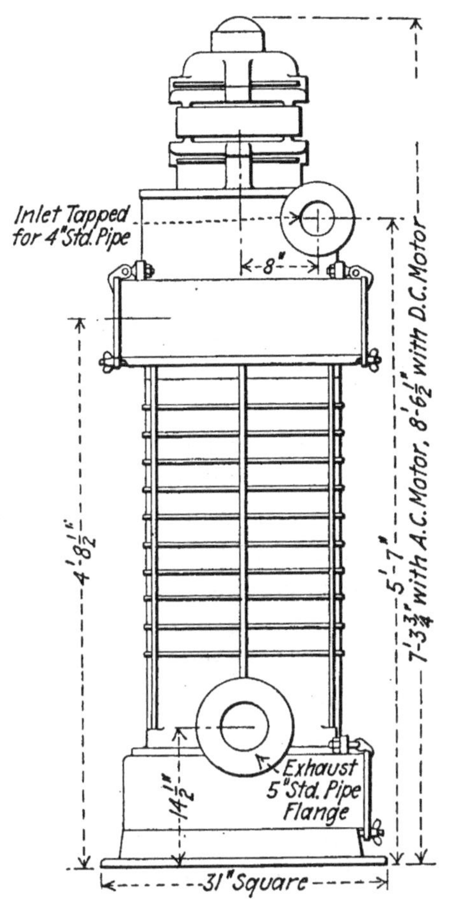

FIG. 10. LATE TYPE OF SPENCER VACUUM CLEANING MACHINE, OPERATED BY MULTI-STAGE TURBINE BLOWER.

cleaning slots varying from 10 in. by ¾ in. to 20 in. by ¼ in. were used, and a very complete line of swivel joints for connecting the renovators and the hose to the handles was developed. This system was operated at 5 in. vacuum, which was much lower than that used by any other system, 15 in. being standard at that time, and a much larger volume of air was exhausted under certain conditions than was possible with any of the then existing systems. Owing to the large volume of air exhausted and to the large size of the renovators, hose and pipe lines, larger articles could be picked up than was possible with any of the existing systems. A great deal of weight was attached to this condition by the manufacturers, a favorite stunt being to pick up nails, washers, waste, small pieces of paper and even pea coal from a floor and finally to pick up a quantity of flour which had first been carefully arranged for the demonstration.

This invasion of the vacuum cleaning field was considered by the established manufacturers as a freak and the apparatus was christened "the tin machine." Whenever it was installed in competition with other forms of cleaning systems, the daily question asked by its competitors was, "Has the tin machine fallen apart?" However, the tin machine did not fall apart, but held its own with the other systems, even in its crude and inefficient state. Finding that the construction he had adopted was too flimsy and subject to abnormal leakage, Mr. Spencer developed a new form of machine, using cast-iron casing and welded fan wheels and adopted standard pipe and fittings. He also brought out a line of sheet metal tools and on the whole perfected a satisfactory cleaning system. One of his machines of a later type is illustrated in Fig. 10.

Separators Emptying to Sewer by Air Pressure.—A new form of vacuum cleaning system was introduced by Mr. Moorhead, of San Francisco, who used an inrush type of renovator having an inlet for air on each side of the cleaning slot.

The separator used with this system was a wet separator and contained a screen cleaned by a rotary brush into which all the dust contained in the air lodged. The pump used with this system was generally of the piston type, fitted with a single rotary valve, so connected to the valve stem that it could be rotated thereon and the machine changed from a vacuum pump to an air compressor in order that the contents of the separators might be discharged into the sewer by air pressure when it was desired to empty same.

This system was marketed by the Sanitary Dust Removal Company, of San Francisco, and, later, was taken over by the American Rotary Valve Company, of Chicago, which is now marketing same. It eliminates the manual handling of the dust at any stage of its removal, a feature which is made much of by its manufacturers, but one which is likely to cause some trouble for the sewerage system if care is not exercised.

Machines Using Root Blowers as Vacuum Producers.—The use of a Root type of rotary pump as a vacuum producer was first undertaken by the Foster and Glidden Engineering Company, of Buffalo, which marketed the Acme system about 1907, the same company having previously built a simliar system for the removal of grain from steam barges. The other features of this system did not differ materially from those already on the market.

Being familiar with the various uses to which this type of vacuum pump had been adapted, the principal one being the operation of pneumatic tube systems, the author suggested the use of this type of vacuum producer about two years previous to its introduction and was advised by one manufacturer that such a type of pump was not suitable for vacuum cleaning. The fallacy of this statement will be brought out in detail in a later chapter.

The type of vacuum producer just described has been adopted in many makes of vacuum cleaners, including the Hope, Connellsville, Arco, and, lately, in the American Rotary Valve Company's smaller systems.

During the past four years a score or more of new stationary vacuum cleaning systems have been introduced, among which are the Palm, a modification of the Dunn-Locke system; the Tuec, a turbine cleaner; the Water Witch, which uses a water-operated turbine as a vacuum producer, and the Hydraulic, with water-operated ejector. At the same time a hundred or more portable vacuum cleaners have been marketed. These are of almost every conceivable type and form and are operated by hand, electricity, and water power. Among them will be found machines which are good, bad and indifferent, the efficiency and economy of which will be discussed in a later chapter.

This nearly universal invasion of the vacuum cleaner field by anybody and everybody looking for a good selling article, establishes the fact that the vacuum cleaner is not a fad or fancy, but has become almost a household necessity and has led large corporations to take it up as a branch of their business. First, the Sanitary Devices Manufacturing Company and the Vacuum Cleaner Company, the pioneers in the field, after a legal battle of years, consolidated with a view of driving their competitors from the field as infringers of the patents controlled by the two organizations. The result of this was the licensing of other companies. In an attempt to control the sale of their type of apparatus notice was served on all users of other types of vacuum cleaners that they were liable to prosecution for using infringing apparatus.

Later, the McCrum-Howell Company, a manufacturer of heating boilers and radiators, secured control of the products of the American Air Cleaning Company and the Vacuum Cleaner Company and sold these machines to the trade for installation by the plumbers and steam fitters. The McCrum-Howell Company has been succeeded by the Richmond Radiator Company, which is handling these vacuum cleaning machines.

Shortly afterwards, the United States Radiator Corporation secured control of the Invincible and the Connellsville systems, and, lastly, the American Radiator Company secured the Wand system.

Thus we see that vacuum cleaning seems to be virtually in the control of the manufacturers of heating apparatus, who are also among the largest corporations in this country and well able to control the future of this business to their liking.

As to the future of vacuum cleaning the author considers that it is at present, like the automobile, at the height of its career, and also, like the automobile, that it is a useful appliance to mankind and that it has its proper place as a part of the mechanical equipment of our modern buildings.

As to the type of vacuum cleaner of the future, the author believes that these appliances will become standardized, just as all other useful appliances have been, and that the form that it will then take will be a survival of the fittest. What that form may resemble the reader may more readily judge when he has completed the reading of this book.

CHAPTER II.

Requirements of an Ideal Vacuum Cleaning System.

Before a comparison of the relative merits of any line of appliances, used for any one purpose, can be intelligently made, one must have either some form of that apparatus which we consider as a standard for comparison that we may rate all others as inferior or superior thereto, or else an ideal of a perfect system must be assumed, and the measures with which each of the various appliances approaches the requirements of the ideal will establish their relative merits.

The author has elected to use the latter method in comparing the various systems of vacuum cleaning, and it is necessary, therefore, to first determine what are the requirements we shall impose on the ideal system.

An ideal vacuum cleaning system would be one which, when installed in any building, will displace all appliances used for dry cleaning in the semi-annual renovating or house cleaning, the weekly cleaning or Friday sweeping and the daily supplemental cleaning. If our system be truly an ideal one, the premises should never become so dirty as to require any semi-annual cleaning at all, and, if the daily cleaning be anyway thorough, there need be no weekly cleaning. This latter condition may be governed by the will of the housekeeper or janitor.

The compressed air cleaners first introduced were intended for use only at the semi-annual cleaning and they were in reality carpet renovators, which were assumed as imparting to the carpets all the beneficial results that could be obtained by taking them up and sending them to a carpet-cleaning establishment, with the advantage over this latter method, that the labor of removal and replacement of the carpets was rendered unnecessary, but with the disadvantage that all the germ-laden air, used as a means of cleaning the carpets, was blown back into the apartment, leaving the germs in their former abode.

This disadvantage, however, is partly offset by the fact that while the majority of the grms in one's own carpet are blown out at the carpet cleaners, a mixed company of germs from your neighbors' and others' carpets, which may be in the tumbling barrel at the same time with your own, are returned to you with your carpet.

Neither of these conditions is ideal and we will expect our ideal cleaner to completely remove from the premises, not only the dust and dirt, but also the germ-laden air which is used as a means of conveying this dirt.

For replacing the weekly and the daily cleaning, these earlier renovators were not suitable, as in order to use same the furniture must all be removed from the apartment.

To accomplish this daily and weekly cleaning, the ideal vacuum cleaner must replace the broom and dust pan, and their inseparable companion, the duster, and must also supersede that time-honored mechanical cleaner, the carpet sweeper.

The reader will doubtless consider that in making this statement the author is asking the vacuum cleaner to perform much more than it is usually called on to do. However, we are now discussing an ideal system, and the above requirements are not absolutely beyond what can be accomplished by some of the cleaning systems now on the market.

To accomplish this requirement the ideal cleaner must pick up everything likely to be found on the floor which cannot be readily picked up by hand. The character of this material will vary greatly according to the uses of the apartment cleaned. In residences and offices, where carpets or rugs are in use, cigar stumps and matches are usually deposited in cuspidors and small pieces of paper in waste baskets, consequently there should be nothing but dust to be removed from a residence and, perhaps, mud and sand from the shoes of the many visitors, in addition to the dust in an office.

However, there are special conditions likely to be met in many cases; sewing rooms will be littered with basting threads and scraps of cloth; department stores, with a great quantity of pins; banking rooms with bands and large-sized bank pins; all of which increase the requirements of the ideal system. A cleaner which is perfectly adapted to one sort of apartment will be entirely unsuited for another, and the ideal cleaner will be one which can be readily adapted to all conditions likely to be met in the building in which it is installed.

The ideal cleaner must be able to accomplish the above stated requirements without the necessity of moving heavy pieces of furniture out of or about the apartment; that is, it must be capable of being efficiently operated under beds, tables and chairs, around the legs of other heavy furniture, behind bookcases, pianos, cabinets, etc., over curtains, draperies and hangings, over walls, behind pictures and over mouldings and carved ornaments, all without injury to any of the furniture or fittings of the apartment, and with the least expenditure of energy by the operator.

These conditions should be met with the fewest possible number of cleaning appliances, none of which should be provided with small attachments liable to be lost or misplaced, and all parts of the system, which must necessarily be moved about, either before, after or during the cleaning operation, should be of minimum weight and bulk, but of rugged and lasting construction.

The ideal vacuum cleaner should be of such proportions and provided with ample motive power to clean rapidly and effectively.

For use in an office building the cleaner should be able to thoroughly clean an average-sized office, including floor, walls, furniture and fittings in from 10 to 15 minutes, and for residence work, should be of sufficient capacity to clean an apartment, including floor, walls, curtains, draperies, pictures and furniture in not exceeding 30 minutes.

The ideal system should be so arranged that any apartment in the building can be cleaned with the least possible disturbance and without affecting the use of any other apartment, excepting perhaps, the corridors or hallways.

In large offices, drafting rooms and similar apartments, it may become necessary to clean same while they are occupied; therefore, our ideal system must be practically noiseless in operation and must offer the least possible obstruction to the proper use of the room by its regular occupants.

Necessity and Proper Location of Stationary Parts.—To be of sufficient power to do rapid cleaning and in order to remove from the building all dust and germ-laden air, the cleaning system must necessarily contain some stationary parts. The motive power can generally be confined to these stationary parts, and must, in such cases, be located within the building to be cleaned. Therefore, it should operate with the minimum of noise and vibration.

Machines located in office or other large buildings, containing elevators or other complicated apparatus requiring skilled attendance, which are provided with complicated control and with other attachments, are not objectionable, and in such cases simplicity should give way to efficiency, but unnecessary complications should be avoided.

In residences and other small buildings, where the vacuum cleaner is likely to be the only machinery installed, the system must be one which requires the minimum attention and must be capable of being started and stopped by any person of average ability, without the necessity of going to the point where the machine is located.

The power consumption of the ideal system should be a minimum to accomplish satisfactory results and should be, as nearly as possible, directly proportioned to the amount of cleaning being done. This requirement is most important in hotels, where some cleaning is likely to be done at all hours, day and night. In other words, vacuum must be "on tap" and as readily attainable at any point in the building as your water or electric light. In office buildings, where a schedule of cleaning hours is fixed, and in residences where cleaning hours are few and the capacity of the plant is rarely more than could be attended to by one operator, this requirement is not of as great importance.

Lastly, our ideal system, from the standpoint of the purchaser, must be of such rugged construction, as will enable it to operate efficiently for, at least, ten years and its mechanical details such that it will operate continuously, without expert attention, and that the annual expense for repairs during the life of the machine will not exceed 5% of the first cost of the system.

CHAPTER III.

THE CARPET RENOVATOR.

In undertaking the comparison of a number of different makes of any appliance, in order to determine the good and bad points in each, where the apparatus is composed of a number of separate and distinct parts, each having its proper function, which they must perform in order to make the whole apparatus effective, as in a vacuum cleaning system, it becomes necessary to isolate temporarily each part and consider its action, first, as a unit working under the most favorable conditions, and, second, as a component part of the whole apparatus in order to determine where the weak points in any system occur and what modifications are necessary in the various parts of the apparatus to make some vital part of the whole more effective. It is further necessary to determine what are the vital parts of the system in order that the other parts may be accommodated to the effective action of that part.

Four Important Parts of Vacuum Cleaning System.—In analyzing a vacuum cleaning system it naturally divides itself into four parts, viz.: the cleaning tool or renovator, the air-conveying system or hose and pipe lines, the separators or other means of disposal of the material picked up, and the vacuum producer.

The author considers that the renovator is the most important part of the system and that the other parts should be made of such proportions and with such physical characteristics as will produce the proper conditions at the renovator to permit it to perform its functions in the most effective manner.

As the vacuum cleaning system must be capable of cleaning surfaces of a widely variable character many forms of renovators are necessary. Of the various surfaces cleaned the author considers that carpets and rugs comprise the most important, as well as the most difficult to clean effectively, so that the carpet renovator will be considered first.

The Straight Vacuum Tool.—Various forms of carpet renovators have been and are in use by manufacturers of vacuum cleaning systems. The first type of renovator to be considered is that having a cleaning slot not over 12 in. long, with its edges parellel throughout its length, and not over ⅜ in. wide, with a face in contact with the carpet not over ⅜ in. wide on each side of the slot. This form of renovator is illustrated in Fig. 11 and is designated by the writer as Type A. The first

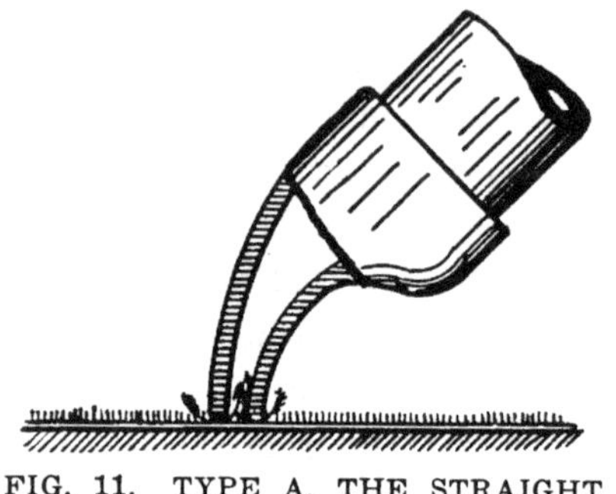

FIG. 11. TYPE A, THE STRAIGHT VACUUM TOOL.

FIG. 12. TYPE B, WITH WIDE SLOT AND WIDE BEARING SURFACE.

of these renovators was introduced by Mr. Kenney and, as finally adopted by him, was 12½ in. long, with ⅞-in. face and with a cleaning slot 11½ in. long and 5/32 in. wide. This form of cleaner was termed the "straight vacuum tool" and is used today by many manufacturers. Slight modifications in its form and dimensions were made in some cases, as in the one manufactured by the American Air Cleaning Company. In the one used in all tests by the writer on type A renovators, the slot was reduced to 10 in. long and ⅛ in. wide and the face of the renovator was slightly rounded at the outer edges, leaving very little surface in contact with the carpet.

A renovator of this type is easily operated over any carpet even when a considerable degree of vacuum exists within the renovator itself. It has met with favor when used with the piston type of vacuum pump without vacuum control, as was the case with the earlier systems. However, when a very high degree of vacuum occurs within the renovator it has a tendency to pull the nap from the pile of the carpet.

Soon after the introduction of this form of renovator, some users of same, particularly in San Francisco, complained that while the renovator effectively removed the dust from carpets it failed to pick up matches and other small articles and preliminary or subsequent cleaning was necessary in order to remove such litter.

To overcome this difficulty Mr. Kenney increased the width of the cleaning slot to nearly ½ in., with the result that when a high degree of vacuum existed within the renovator, which often occurred where no vacuum control was used, it stuck to the carpet, rendering its operation difficult and, at the same time, doing great damage to the carpet. Hence, its use with the piston type of vacuum pumps was abandoned.

Mr. Kenney then modified this wide slot renovator by making the face of same much wider, thus having more surface in contact with the carpet on each side of the slot, preventing the renovator from sinking into the nap of the carpet. This type of renovator is illustrated in Fig. 12 and has been designated as Type B. While not as destructive to the carpets, when a high degree of vacuum existed under the same, it still pushed hard and was not as rapid a cleaner as the narrow-lipped Type A renovator.

Renovator with Auxiliary Slot Open to Atmosphere.— The renovator introduced by the Sanitary Devices Manufacturing Company differed widely from the former types in that it was provided with an auxiliary slot, open to the atmosphere through the top of the renovator, which communicated with the slot open to the vacuum by a space of 1/32-in. under the partition separating the slots. The cleaning slot was made 5/16-in. wide and the face of the renovator was made 2-in. wide, which gave a contact of 13/32-in. in front of the inrush slot and 21/32-in. in the rear of the cleaning slot. This form of renovator is illustrated in Fig. 13 and is designated as Type C.

The auxiliary slot or vacuum breaker permitted air to enter the cleaning slot even when the renovator was placed on a surface plate, and, owing to this feature, a high degree of vacuum never existed within the renovator. It was always easy to operate and did not damage the carpet. Owing to the wide slot, articles of considerable size could be picked up, and there was always an abundance of air passing through the renovator to produce a velocity in the hose and pipe lines sufficient to carry any heavy articles picked up.

The vacuum producer, control apparatus and the proportions of the hose and piping used at that time made the degree of vacuum in the renovator a function of the quantity of air passing, with wide limits of variation under existing conditions, and this form of renovator is practically the only one which will do effective cleaning, including the picking up of litter, without undue wear on carpets, when used with a system having the above-stated characteristics. This renovator, however is not without its faults. Owing to the wide surface in contact with the carpet, a considerable degree of vacuum is necessary in order that any air shall enter the renovator under

FIG. 13. TYPE C, WITH AUXILIARY SLOT, OPEN TO ATMOSPHERE.

FIG. 14. TYPE D, WITH TWO CLEANING SLOTS.

the faces of same and, as the air entering the inrush slot prevents the formation of such vacuum within the renovator, very little air enters the renovator between its face and the carpet. When the renovator is operated on a carpet having a glue-sized back, no air enters through the carpet, therefore all air entering the renovator must come through the inrush slot and under the partition separating same from the cleaning slot. Under these conditions only one side of the vacuum slot is effective and this effective side is raised above the surface of the carpet.

When operated on an ingrain or other loose-fabric carpet, much air enters through the fabric of the carpet, due to the wide cleaning and inrush slots, in addition to the quantity of air entering through the inrush slot, making this renovator, when operating under these conditions, use an unnecessary amount of air. Apparently, this renovator has been designed to prevent the formation of any great degree of vacuum under same and such a design has resulted in a greater volume of air at a lower vacuum passing through than through renovators of other types.

This property of the renovator raises the question whether the quantity of air or the degree of vacuum in the renovator is most essential for the removal of dirt from carpets. Tests made by Mr. S. A. Reeve, consulting engineer for the Vacuum Cleaner Company, with this type of renovator, with the inrush open and repeated with the inrush closed, disclose the fact that it does more effective cleaning with its inrush closed, while the volume of air passing is considerably less with the inrush closed. The degree of vacuum was greater, which tends to indicate that the vacuum within the renovator is the most important factor.

An extract from the affidavits of Mr. Reeve in one of the numerous patent suits will show his explanation of this phenomenon: "If we examine more closely into the actual process whereby such a sweeper succeeds in extracting dust from carpets, etc., it will appear that the actual cleaning is effected at the periphery of the slot in the lower surface of the sweeper. It is accomplished chiefly by the development of local changes of air pressure at the lips defining this slot, incidentally to the movement of the tool over the carpet. These changes cause the air occupying the interstices between the dust particles to expand suddenly, thus 'raising the dust.' To a lesser degree, the scouring is effected by highly localized air currents of considerable velocity, engendered where the tool comes in contact with the carpet. These air currents pick up the dust which has already been expanded or raised by pressure change. They will be of higher velocity, and therefore more effective, the better the contact of the tool with the carpet. The same is true of the pressure changes.

"All this action depends for its intensity, speed and effectiveness, not on the vacuum existing at the pump or in the separators, but upon the vacuum prevailing within the sweeper head itself."

Renovator with Two Cleaning Slots.— Another form of renovator was introduced by the Blaisdell Machinery Company which contained two cleaning slots each 3/16-in. wide and 12-in. long, separated by a partition ¼-in. wide in contact with the surface of the carpet, as indicated in Fig. 14 (Type D). While this form of renovator has a greater area of cleaning slot than Type A, its individual cleaning slots are no wider; therefore, it cannot pick up anything larger than can be picked up by Type A. As no air can enter under the partition it can do no more effective work as a dust remover when operated on a carpet with a glue-sized back and its only advantage over a cleaner of Type A is that when operated on a loose-fabric carpet more air can pass through the fabric into the cleaning slot, thus giving a greater variation in the quantity of air exhausted when operated on carpets of different texture, a condition which is undesirable when used with a system having characteristics previously described.

Tests of this type of renovator, made by Mr. Reeve, are given later in this chapter.

Renovator with Inrush Slots on Each Side.—Another form of renovator, introduced by Mr. Moorhead, is illustrated in Fig. 15 (Type E). This is a modification of Type A in that an inrush slot is provided on each side of the vacuum slot, these inrushes being hinged members which form the sides of the cleaning slot. This cleaner has the advantage over Type C renovator in that it can take air from either side, but in action it takes air from but one side at any time. Its inrush will not become entirely clogged, but its mechanically-moving parts in contact with the dust and lint picked up will easily become inoperative and are as like as not to become caught wide open when the air entering the cleaner will not come into intimate contact wtih the carpet. In that event, its cleaning efficiency will be greatly reduced. The author has not had an opportunity to make any comparative tests of this form of renovator.

When Mr. Spencer introduced the centrifugal fan as a vacuum producer, he also brought out a series of carpet renovators of various forms and sizes. One had a cleaning slot ¾-in. wide and 10-in. long, another a slot 15-in. long, ¼-in. wide at its end, increasing to ¾-in. at the center. Another had a slot 20-in. long and ⅜-in. wide, and finally he adopted a tool with a cleaning slot 15-in. long and ½-in. wide throughout its length. This is merely the re-entrance into the field of the wide-slot tool first used by Mr. Kenney and its successful operation depends on its use with a vacuum producer of such characteristics and a hose and pipe line of such proportions that practically a constant vacuum is maintained within the renovator, regardless of the quantity of air passing through the tool. The latest form of this renovator, as used by Mr. Spencer, is illustrated in Fig. 16. At the time that the writer made tests on renovators of this make, the majority of the tests were made with a renovator having a cleaning slot 10-in. long and ¾-in.

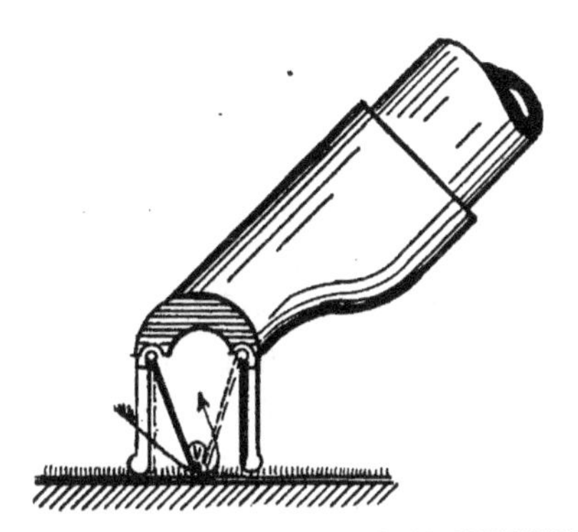

FIG. 15. TYPE E, WITH INRUSH SLOT ON EACH SIDE OF VACUUM SLOT.

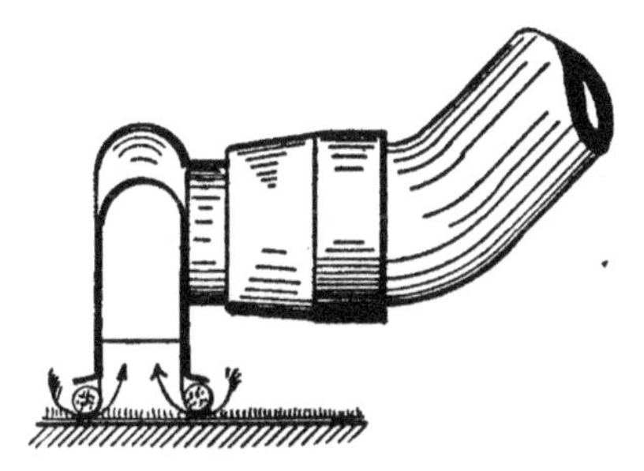

FIG. 16. TYPE F, AN EXAGGERATED FORM OF TYPE B.

wide. This renovator is designated as Type F, while the 15-in. x ¼-in. to ¾-in. slot is designated as Type F^1.

About seven years ago the Supervising Architect of the United States Treasury Department gave consideration to the use of a carpet cleaning test to determine the acceptability of any vacuum cleaning system which might be installed in any of the buildings under his control. The author was instructed to make a series of tests of carpet renovators, with a view of determining: (1) the feasibility of using a carpet cleaning test to determine the merits of a vacuum cleaning system; (2) to fix the requirements to be incorporated in a specification where the acceptance of the system was dependent on a satisfactory carpet cleaning test, to be made at the building after the completion of the installation; (3) to determine what requirements, other than a cleaning test, would be necessary to obtain a first-class cleaning system.

The record of many such tests was shown to the author, shortly before he began making tests. These purported to have been made by Prof. Miller at the Massachusetts Institute of Technology, with a pump furnished by the Sanitary Devices Manufacturing Company, in which the efficiency of the inrush type of renovator (Type C) and the straight vacuum renovator (Type A) was compared. The results of these tests, as given in a brief resumé, which was distributed by the Sanitary Devices Manufacturing Co., indicated that the Type C renovator was the more rapid and efficient cleaner.

The author learned that these tests were made by the undergraduate students as a part of the regular laboratory work, and that later a series of tests was made as the basis of a thesis by Messrs. Paterson and Phelps in 1906, using the above-described apparatus. The following year another series of tests was made by Mr. Stewart R. Miller, as the basis of an undergraduate thesis, in which the efficiencies of the piston pump and inrush sweeper of the Sanitary Devices Manufacturing Co. were compared with those of the steam aspirator and straight vacuum renovator of the American Air Cleaner Company. A copy of this thesis was furnished the author by the Sanitary Devices Manufacturing Company shortly after the completion of the tests made by the author.

The relative efficiency of the two types of renovators reported by these tests differed widely in each case, an occurrence which is liable to happen where undergraduate students are engaged in such work. They were, therefore, considered as of doubtful reliability.

The author could find no record of any tests made by anyone of longer experience and, indeed, these were the only tests of which he could find any record.

As the author desired to specify a cleaning test which could be readily repeated at the building in which the cleaning system was installed, which building was likely to be located in any part of the United States, no exhaustive laboratory methods were desired or attempted. As the building was likely to be located in a city where no other vacuum cleaning systems were then installed and in a new building in which no dirty carpets were available, and as it was not desirable to have the contractor furnish the material for the test, it was considered necessary to use some material in soiling carpets which would be readily obtainable anywhere, which could be readily brought to a standard, and which, when worked into the carpets in a reasonable length of time, would be as difficult to remove as the dirt found in the average dirty carpet.

Tests on Dirty Carpets.—As no tests of cleaning an actually dirty carpet were on record, quicksand having been used in the Institute of Technology tests, it was necessary to first clean some carpets that had been soiled in actual daily service in order to obtain a standard with which to compare the results in removing various substances, which it was intended to try as a substitute for dirt. A carpet which had been in actual use for a number of years on the floors of the old United States Mint building, in Philadelphia, and receiving the ordinary amount of cleaning, was procured. This was a Brussels carpet with a glue-sized back, containing about 20 sq. yds. It was divided into three approximately equal parts.

An indicator was attached to the vacuum pump for taking air measurements, and it was found that there was considerable leakage of air into the system through the connections to the separators and at other points, therefore the pump was operated with 22 in. of vacuum in the separator and a card taken with all outlets closed and the amount of leakage noted. During the tests this degree of vacuum was always maintained in the separators and pipe lines and the vacuum in the renovator was varied throughout the tests by throttling the hose cock. This manner of making tests gave a practically constant leakage which was deducted from the quantities shown by the indicator cards taken with the renovators in operation.

As the writer had already made many tests of the efficiency of various types of vacuum pumps as air movers under various degrees of vacuum, and as the capacity of the pump available was far in excess of that required to operate one renovator, no attempt to obtain the efficiency of the plant as a unit was made. Instead, the vacuum at the hose cock was adjusted until the degree obtained was what the writer had found to be within the limit obtained in practice. The resulting vacuum at the renovator was then noted.

Each piece of carpet was cleaned during six periods of one minute each, using a different vacuum at the tool for each piece of carpet. The carpets were weighed at the beginning of the test and after each one-minute period. At the conclusion of these tests each carpet was cleaned until no change of weight occurred after two minutes' cleaning. They were then considered as being 100% clean and this standard was made a basis for computing the percentage of dirt removal. A renovator of Type C was used in these tests.

Shortly afterward a similar test was made on a dirty carpet of 4.6 sq. yds. area, using a renovator of Type F. This carpet was also a Brussels, with glue-sized back, which had been in use in the shoe department of a large department store in Hartford. These carpets contained approximately 2 oz. of dust per square yard, none of which was visible on the surface, and they were probably as clean as the average carpet after being

TABLE 1.

CLEANING TESTS OF DIRTY CARPETS.

Type of Renovator.	A		C			F
Vacuum in renovator, in. Hg	2	4½	1	2½	4	3½
Air exhausted, cu. ft. per min	16	27	24	37	44	59
Material removed, per cent. of total, 1 min	50	60	37	39	47	35
Material removed, per cent. of total, 2 min	72	81	52	59	63	55
Material removed, per cent. of total, 3 min	85	90	59	66	71	69
Material removed, per cent. of total, 4 min	90	95	61	72	83	77
Material removed, per cent. of total, 5 min	93	98	66	75	87	84
Material removed, per cent. of total, 6 min	95	100	67	82	90	89
H. P. per ounce dust	0.037	0.147	0.045	0.116	0.252	0.261
Ounces dust per minute	1.9	2.0	1.34	1.64	1.8	1.78
H. P. at renovator	0.07	0.29	0.06	0.19	0.45	0.475

gone over with a carpet sweeper or after a light application of a broom.

As the sizes of the carpets used in making the tests were not always the same, allowance has been made for this variation by using, in the case of Type F renovator, instead of the true time, a calculated time which allows each renovator the same time for cleaning 1 sq. yd. of carpet. For instance, in the case of the small carpet cleaned with Type F renovator, an interval of $60 \times 4.6 \div 6$, or 46 seconds, was taken as equal to one minute's cleaning of the carpet with types A and C renovators. Such interval is stated and plotted as one minute in the table opposite, which gives the results of cleaning dirty carpets with the three types of renovators.

Type A Renovator Most Efficient on Dirty Carpets.— The results of the tests of the three types of renovators, each

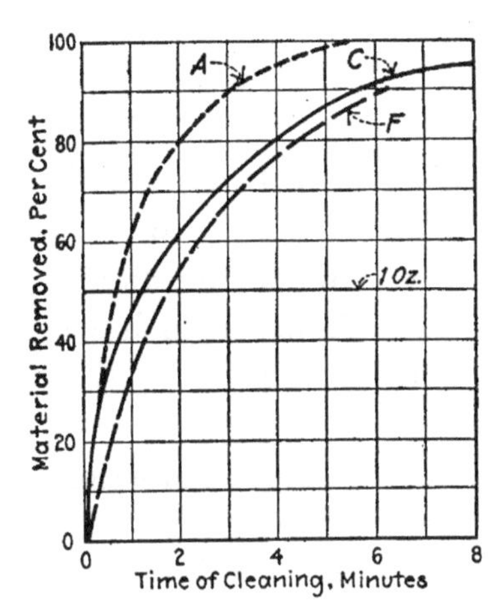

FIG. 17. TESTS OF THREE RENOVATORS ON DIRTY CARPETS.

when it was operated with the highest vacuum under the renovator, are plotted in Fig. 17 in order that a ready comparison may be made. This curve indicates that Type A renovator does more effective cleaning in less time than either of the other two types tested.

Referring to the second line of the table, which gives the degree of vacuum obtained in the renovator during the tests, it will be noted that the highest vacuum attained with each type of renovator is practically the same. This degree of vacuum was obtained with the average vacuum at the hose cock, using 100 ft. of hose in each case, and corresponds to that obtained in the commercial operation of each of the renovators with the vacuum producers ordinarily used, which was 15 in. in the case of Type C, 10 in. in case of Type A, and 5 in. in case of Type F, the hose being the size used by each of the systems as marketed.

The third line, which shows the cubic feet of free air per minute passing the renovator, indicates that Type A renovator requires much less air at the same degree of vacuum than either of the other types to do better work.

From the readings in these two lines the horse power required at the renovator, to move the air that passes same is obtained with 100% efficiency adiabatic compression. The results are tabulated in the ninth line of the table.

This indicates that Type A renovator does more effective work with about 50% of the power required by either of the other types of renovators.

The tenth line gives the rate of cleaning and again shows Type A renovator to be the most rapid cleaner.

The eleventh line gives the horse power required at the renovator when in operation, from which it will be seen that effective cleaning cannot be accomplished with less than ¼ H. P. at the renovator.

Attention is called to the great reduction in power in case of Type A renovator when the vacuum at the tool is reduced from 4½ in. to 2 in. and to the small reduction in the efficiency which results from this great reduction in power. This is not the case with the Type C renovator, where there is a considerable reduction in the already low efficiency with each reduction in the vacuum. This characteristic of Type A renovator is discussed further on in the chapter on hose.

Tests of Carpets "Artificially" Soiled.—Having determined the efficiency of the various types of renovators when operated on dirty carpets, the author then attempted to find some substance easily obtained anywhere which could be used as a substitute for actual dirt, and which would give approximately equal results with these obtained on dirty carpets.

A test of this character was made by the author some time previous to the tests of dirty carpets and was made on a Wilton velvet rug of about 12 sq. yds. area. The material spread on same was ordinary wheat flour, as used in demonstrations, 3 lbs. of which were placed on the rug and rubbed in with sticks of wood as well as possible and the rug cleaned for three minutes, using a Type A renovator attached to the separator with 50 ft. of 1-in. diameter hose. The results were as follows:

Vacuum at Separator, Ins. Mercury.	Per Cent. Dirt Removed.
5	95
10	98
15	98

The vacuum at the renovator was not measured at the time of making this test and its amount is not exactly known, but further tests with this type of renovator under nearly the same conditions gave the following results:

Vacuum at Hose Cock, Ins. Mercury.	Vacuum in Renovator, Ins. Mercury.
5	3
10	6½
15	9

and it is probable that the vacuum at the renovator during these tests was approximately the same.

Comparison of the results of this test, in which 4 sq. yds. of carpet were cleaned per minute, wtih those of the tests of dirty carpets, in which only 1 sq. yd. was cleaned per minute, indicates that wheat flour is not a suitable substitute for dirt in making a carpet cleaning test.

The author, believing that flour is of sufficient fineness, but not of sufficient weight, tried Portland cement, which is very heavy and at the same time exceedingly fine, as a substitute for dirt in soiling carpets. The same carpet that had been cleaned in Philadelphia was used and 6½ oz. of cement was worked into the same. It was then cleaned with a Type C renovator, with a vacuum of 2½ in. hg. at the renovator and 95% of the cement was removed in two minutes' cleaning, as against 59% of the dirt in the carpet when received.

Ordinary dirt, taken from some flower pots which had been

left dry for some time, was then tried with the same carpet, using a Type C renovator aand 1 in. hg. With this arrangement, 71½% of the dirt was removed in two minutes as against 52% of the dirt in the carpet as received.

This dirt was then mixed with water to a thin mud and spread over the carpet and the carpet dried before cleaning. Then 11¼ oz. of this material was worked into 6 sq. yds. of carpet and a Type C renovator removed 100% of this in four minutes' cleaning, with a vacuum of 2½ in. hg. at the tool as against 72% of the dirt in the carpet as received.

The author's ingenuity being about exhausted, he referred to the test of Mr. Stewart R. Miller in which quicksand which would pass a 50-mesh to the inch screen was used, a long-napped Brussels carpet being filled with 5½ oz. per square yard and cleaned with Types A and C renovators.

This test indicated that a nearer approach to the results in cleaning dirty carpets was possible with this substance than with any which the author had tried. The author repeated Mr. Miller's test, using a Type F renovator, 10-in. x ¾-in. cleaning slot, and also a Type F[1] renovator, 15-in. x ¼-in. to ¾-in. cleaning slot. In duplicating these tests the author was associated with Mr. E. L. Wilson, a graduate of the Institute, who

TABLE 2.

CLEANING TESTS OF CARPETS FILLED WITH 5½ OZ. OF QUICKSAND PER SQUARE YARD OF CARPET.

Type of Renovator.	A	C	F	F'
Vacuum in renovator, in. hg.............	4½	4	3½	3½
Air exhausted, cubic feet per minute......	27	44	59	54
Material removed, per cent. of total, 1 min.	60	53	66	53
Material removed, per cent. of total, 2 min.	75	65	83	75
Material removed, per cent. of total, 3 min.	82	74	94	86
Material removed, per cent. of total, 4 min.	87	82	100	94
Material removed, per cent. of total, 5 min.	92	87	—	100
Material removed, per cent. of total, 6 min.	95	93	—	—
H. P. per ounce sand....................	0.09	0.138	0.084	0.109
Ounces sand per minute..................	3.2	3.1	5.3	4.0

was familiar with the methods used by Mr. Miller. With his assistance, the conditions of Mr. Miller's tests were almost exactly duplicated. The results of Mr. Miller's and the author's tests are given in the table opposite, correction being made in the time of cleaning proportional to the size of carpets used, to allow the same time for cleaning 1 sq. yd. of carpet by each renovator.

The results of these tests are shown graphically in Fig. 18. Comparison of these curves with the curves of cleaning dirty carpets (Fig. 17), shows a falling off in the efficiency of cleaning by Type A renovator while there is a gain in the efficiency in cleaning by all of the other types of renovators, Type C

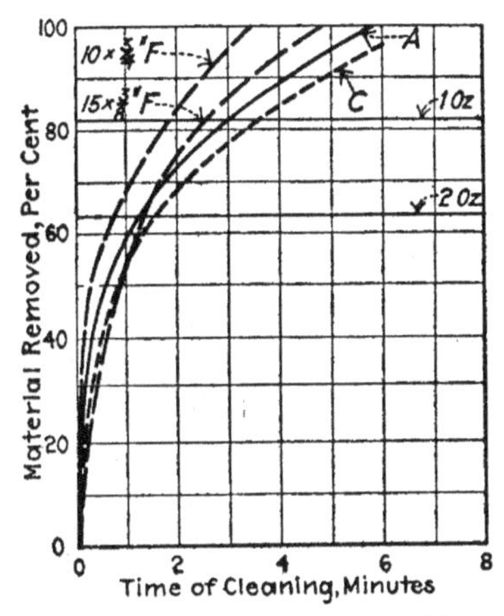

FIG. 18. CLEANING TESTS OF CARPETS FILLED WITH QUICKSAND.

being now nearly as efficient as Type A, while Types F and F[1] renovators are now more efficient than Type A. This result must be due either to the increased quantity of material to be removed, 5½ oz. per square yard in case of the sand as against 2 oz. per square yard in case of the dirt, or else to the change in the character of the material removed, the sand having much sharper surfaces than would be encountered in case of dirt which must necessarily be ground under the feet before it reaches the carpet, or to the longer nap of the carpet.

In order to determine the effect of the increase in the quantity of material on the results, the tests were repeated using

1 oz. of sand per square yard of carpet in each case, omitting the test on Type F[1] renovator.

These tests were made on a glue-sized back, short napped Brussels carpet, using as much sand as could readily be worked out of sight in this carpet. The results of tests are given in the following table:

TABLE 3.

CLEANING TESTS USING 1 OUNCE OF SAND PER SQUARE YARD OF CARPET.

Type of Renovator.	A		C			F
Vacuum in renovator, in. hg........	2	4½	1	2½	4	3½
Air exhausted, cubic feet per min....	16	27	24	37	44	59
Material removed, per cent. of total, 1 min..............................	48	54	45	48	50	50
Material removed, per cent. of total, 2 min..............................	70	87	60	63	65	73
Material removed, per cent. of total, 3 min..............................	91	100	73	75	77	87
Material removed, per cent. of total, 4 min..............................	100	—	76	81	88	100
Material removed, per cent. of total, 5 min..............................	—	—	—	88	97	—
Material removed, per cent. of total, 6 min..............................	—	—	—	92	102	—
H. P. per ounce sand...............	0.047	0.143	0.06	0.195	0.44	0.223
Ounces sand per minute.............	1.5	2.0	—	0.92	1.02	2.11

The results of these tests at the higher vacua are shown graphically in Fig. 19. Comparison of these curves with those obtained when removing sand from a long napped carpet (Fig. 18), shows:

First, a marked increase in the efficiency of Type A renovator, this being slightly better than obtained when cleaning a dirty carpet.

Second, practically no change in the efficiency of Type C renovator.

Third, a small decrease in the efficiency of Type F renovator, which still shows a much higher efficiency than when cleaning dirty carpets.

In order to determine how much, if any, of these changes in the behavior of the renovators was due to the increase in the quantity of material to be removed, the horizontal line, representing 1 oz. of sand remaining in the long-napped carpet, was drawn on Fig. 18 and, using this as a base line, it will be seen that Type A renovator removes this remaining material in three minutes, the same time as was required to remove the same amount from the short-napped carpet. However, the first 4½ oz. of sand have been removed from the long-napped carpet in three minutes, or at a rate 4½ times as fast as the last 1 oz. was removed. This indicates that the narrow slot renovator is capable of handling more material than is likely to be encountered in any dirty carpet and that the apparent decrease in the efficiency of this renovator is not due to the increased quantity of material to be removed.

It will be noted that the Type C renovator removed the last 1 oz. per square yard from the long-napped carpet in the same

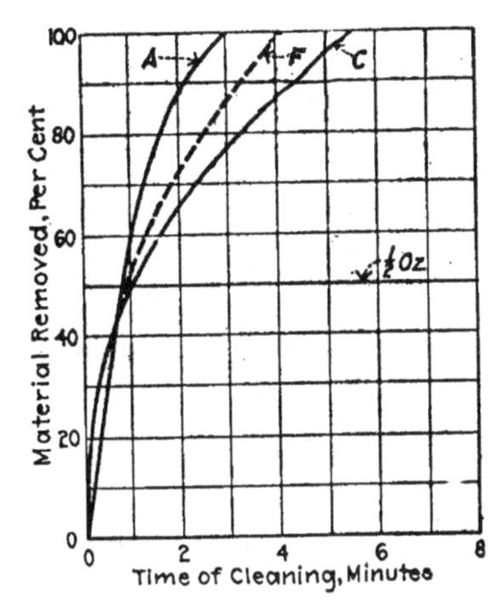

FIG. 19. CLEANING TESTS USING 1 OZ. OF SAND PER SQUARE YARD OF CARPET.

time that was required by Type A renovator, while it needed nearly twice as long to remove this amount of material from the short-napped carpet (Fig. 19). This renovator, however, was slower in removing the first 4½ oz. per square yard.

Type F renovator removed the last 1 oz. per square yard from the long-napped carpet in two minutes, while it required twice this time to remove the same amount from the short-napped carpet. This renovator also removed the first 4½ oz. per square yard from the long-napped carpet in two minutes, while it required three minutes for Type A and 3¾ minutes for Type C renovators to remove the same quantity. It is, therefore, evident that sand is removed more rapidly from a long than from a short-napped carpet when a wide slot renovator is used. The same time is required to remove small quantities of sand from a long or short-napped carpet with a narrow slot.

This phenomenon is probably due to the sand being held in the carpets by the adhesion of its sharp edges to the sides of the nap, this being more pronounced in the case of the long-napped carpet where it is easier to work the material out of sight without grinding it into intimate contact with the pile of the carpet. When the wide-slot renovator passes over the carpet, the carpet is arched up into the slot and the upper ends of the nap separated. The longer the nap or the wider the slot, the greater will be this separation. With the long-napped carpet this separation will at once release the sand, while, in case of the short nap, there is less separation and also more adhesion of the sand to the pile of the carpet, due to the harder grinding necessary to work the material out of sight. Therefore, the wider the cleaning slot used, the faster the sand will be removed, as is evident by comparison of the tests of Types F and F^1 renovators on the long-napped carpet.

With the narrow slot renovator the arching of the carpet under the cleaning slot is negligible and no advantage is gained when using this type of renovator to remove sand from a long-napped carpet. It is also possible that the nap of the carpet may be longer than the width of the cleaning slot, in which case the nap will not snap back to a vertical position when it is under the cleaning slot, but will be pressed down and will impair the action of the renovator. The author considers that the width of the slot should always be greater than the length of the nap of the carpet in order to do effective cleaning.

Shortly after making the above-described tests, the author had occasion to make somewhat similar tests, using a sand-filled carpet, in an attempt to try out a proposed carpet cleaning test intended to be used as a standard for use in specifications for a vacuum cleaning system. When a Wilton carpet was used, it was found that neither Type A or C renovator would fulfill the test requirements, which were within the results obtained in tests already described. Unfortunately a Type F renovator was not available, but the author is of the opinion that it would have done better.

The test was then repeated, using a Brussels carpet and the test requirement was easily met. This discovery led the author to make further tests of carpets of different makes, filled with sand and cleaned under the same conditions which yielded far from uniform or satisfactory results, and the use of a cleaning test, where artificially-soiled carpets are used, was abandoned.

The author is of the opinion that no substance artificially applied to a carpet, other than regular sweepings, will give anything like the same results as will be obtained in actual cleaning. Sand seems to be the only substance which can be worked into the carpet that is nearly as difficult to remove as the actual dirt found in carpets, and, in many cases, this material gives results that are misleading and unfair to some types of renovators. No test which uses a carpet artificially soiled with artificially prepared dirt is considered to be of any value in determining the relative efficiency of various types of carpet renovators.

A series of tests was made by Mr. Sidney A. Reeve consulting engineer, of New York City, in October, 1910, at the works of the Vacuum Cleaner Company, Plainfield, N. J., in which the conditions were such as would give much more uniform results than were possible in the tests made by the author.

In making these tests the renovator was held firmly clamped in any desired position in a wooden carriage rolling upon a straight wooden track. The portion of the carriage supporting the sweeper is attached to the remainder of the carriage by hinges, so that the sweeper is free to seek its own contact with the carpet. The carriage was given a reciprocating motion by its attachment to a large bell crank, which in turn received its motion from the factory shafting. The construction of the bell crank was such that the driving power could be readily thrown in and out of gear at any time.

The carpet was stretched tightly upon a platen which was fitted for movement across the line of motion of the sweeper, along straight guides suitably attached to the floor. The ends of the carpet were first wedged tightly in clamps and the clamps wedged apart so as to stretch the carpet.

The tests consisted in first weighing the carpet, then stretching it upon the platen, then sprinkling thereon a suitable and known weight of dirt taken from the separators of the company's machines, from which the lint and coarse, fibrous material had been sifted and which was thoroughly trodden into the fibres of the carpet, whereupon the sweeper was set in motion for a given number of strokes.

In nearly all cases the tests were repeated upon the same piece of carpet, with the same charge of dirt, by repeatedly placing the carpet in the frame and giving it a further and more extended cleaning.

All tests were corroborated by repetition before being admitted to the records. Every effort was made to have the tests approach the conditions occurring in actual practice, as nearly as possible, and still keep them definite and measurable.

The carpet used was a Wilton, of the standard width of 27 in. and something over a yard long, and the sweeper was given a stroke of 34 in. at the rate of 40 strokes per minute. The sweepers were attached to a 6-ft. tubular handle, 15/16-in. inside diameter, and connected to the separator by 50 ft. of 1-in. diameter hose.

Before making any tests, the piston pump used in the experiments was calibrated by pumping through a rotary meter and the amount of air moved per revolution for each degree of vacuum from open inlet to closed system was carefully determined. In making the tests of various renovators, each renovator was allowed to pass the same amount of air as the others tested in comparison therewith and the vacuum at the renovator and at the separator was allowed to be what was necessary to pass this known amount of air through the renovators. This method is widely different from that used by the author where the degree of vacuum at the renovator head was determined and used as a limiting factor, the quantity of air being allowed to vary as necessary to produce this vacuum.

The results of three series of tests are given in Fig. 20, which shows those obtained with Kenney Type A renovators, having a face 12½ in. x ⅞ in. and a cleaning slot 11½ in. x 5/32 in.

Curve A was made with the angle of the handle such as would give as near as possible a perfect contact of the sweeper with the carpet. Curve B was made with the sweeper handle canted 5° below the proper angle. Curve C was made with the sweeper handle raised approximately 15° above the proper angle. The ordinates represented the amount of dust in the carpet in 40ths of a pound, also reduced by the author to ounces, and the abscissae the number of strokes made by the sweeper.

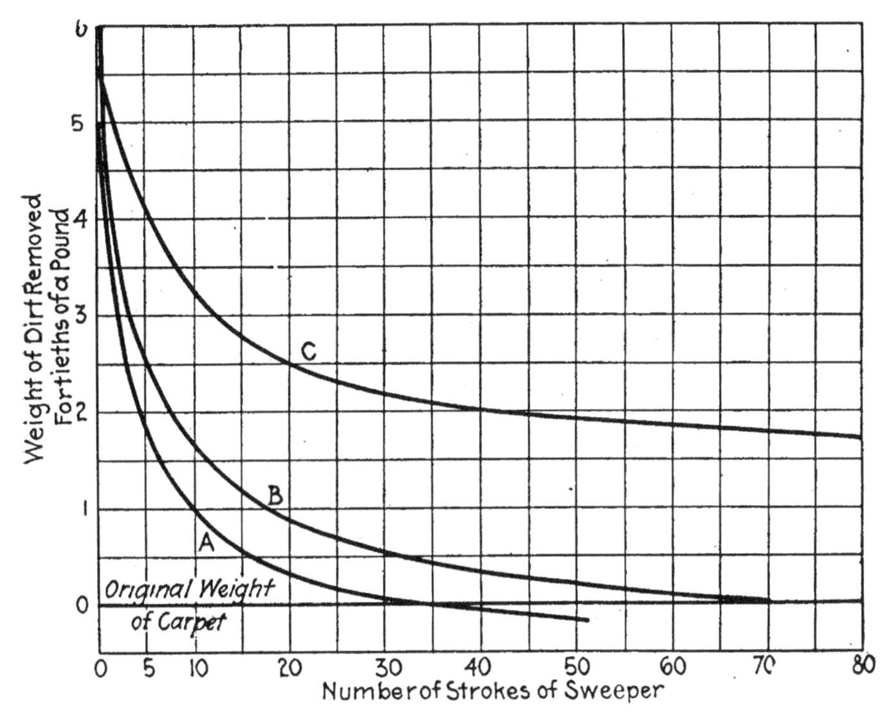

FIG. 20. THREE SERIES OF TESTS WITH KENNEY TYPE A RENOVATORS.

Curves B and C show the loss in efficiency which occurs when the renovator is canted from its proper position on the carpet. This falling off in efficiency will necessarily be greater the wider the face of the renovator, as is shown in further tests by Mr. Reeve, using a Type C renovator, which tests also show that this renovator gives a slightly higher efficiency when operated with the inrush slot stopped, as is shown in Fig. 21.

In this curve the ordinates represent the per cent. of normal dirt, *i. e.*, the amount likely to be found in a dirty carpet, remaining in the carpet at any stage of the cleaning, and the abscissae the number of strokes that have been made by the sweeper. Heavy solid lines represent the results with the inrush open and dotted lines the results with the inrush stopped. The figures on the curve represent the degree to which the handle has been varied from the position giving the best results in cleaning.

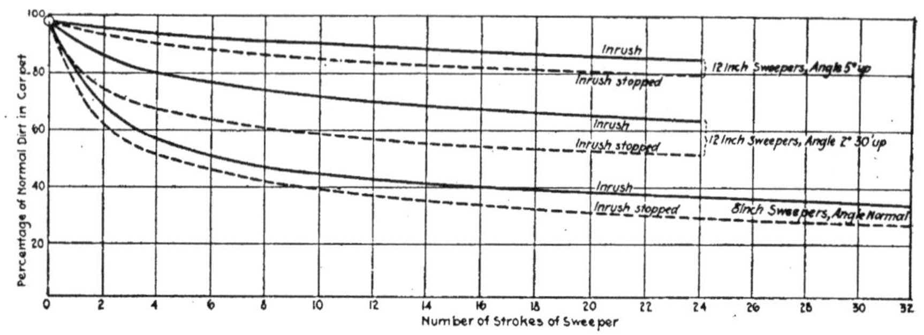

FIG. 21. TESTS BY MR. REEVE, USING TYPE C RENOVATOR.

Fig. 22 shows the results of tests by Mr. Reeve using a renovator of Type D, having a double cleaning slot, and indicate that this type of cleaner is not as efficient as Type A and is affected more by the canting of the handle from the best angle for cleaning.

The above mentioned tests are published through the courtesy of Messrs. Ewing and Ewing, attorneys for the Vacuum Cleaner Company.

Since the method of making these tests is entirely different from that used by the author, a comparison of the results, with any assurance that the same conditions existed in both cases, is impossible. It occurred to the author that a comparison of the results of the tests by Mr. Reeve, using a carpet artificially filled with actual dirt taken from carpets, with the tests made by the author on carpets naturally soiled, would tend to show if equal results could be obtained by a vacuum cleaner by artificially soiling a carpet with dirt taken from another carpet, and in cleaning a carpet naturally soiled.

The author has reduced these results to the same units of time per square yard of carpet cleaned as in the test on the Philadelphia carpet with the small-sized Type A renovator (11-in. x ½-in. face and 10-in. x 3/16-in. cleaning slot). The carpet used by the author contained 6 sq. yds. and was held in cleaning by a weight at each corner, while the carpet used by Mr. Reeve was ¾ yd. wide and cleaned for approximately one yard of its length, the relative size being 1 to 8. The time of cleaning was 6 min. in the author's test which would corre-

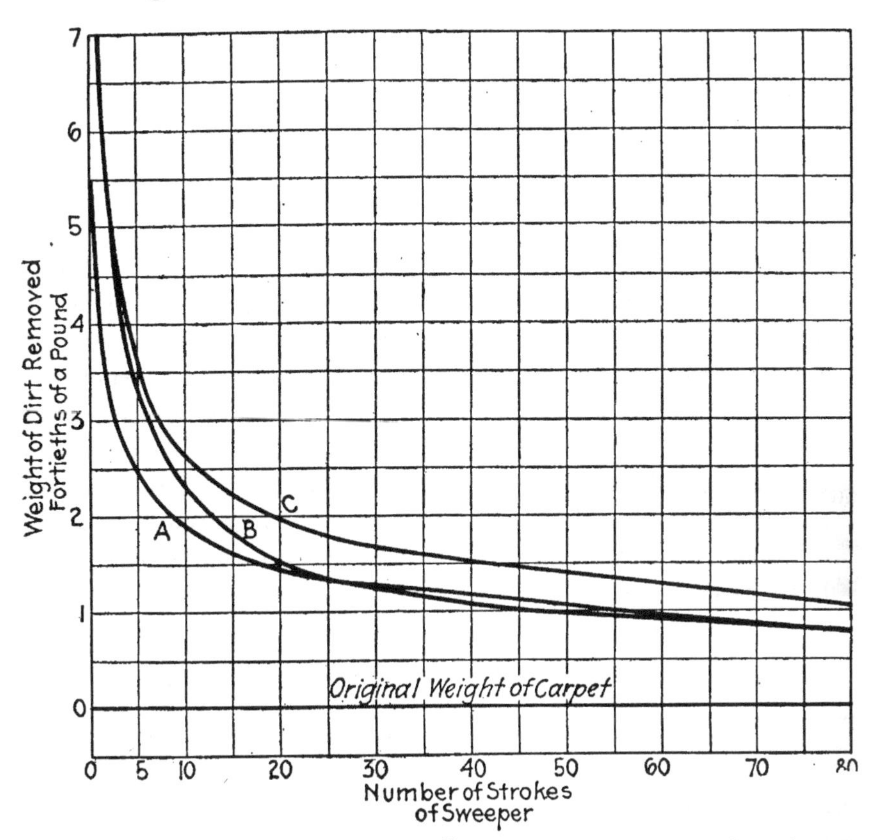

FIG. 22. TESTS BY MR. REEVE, USING TYPE D RENOVATOR.

spond to ¾-min. cleaning in Mr. Reeve's test, or 30 strokes of the sweeper. The total dust in the carpet in Mr. Reeve's test was 5/40 lbs., or 2.66 oz. per square yard, and his test is compared with the author's test with the carpet containing 2 oz. per square yard. Calculation of the per cent. of total dirt removed in each 5 strokes of the sweeper in Mr. Reeve's test, and a comparison of the per cent. of dirt removed in each one minute's test by the author are given below:

TABLE 4.

COMPARISON OF TESTS MADE BY MR. REEVE AND BY THE AUTHOR.

Mr. Reeve's Test.		Author's Test.	
Strokes.	Material removed, per cent. of total.	Minutes.	Material removed, per cent. of total.
5	62	1	60
10	80	2	81
15	89	3	90
20	94	4	95
25	97	5	98
30	99	6	100

The above comparison was made using curve A, Fig. 20, with the sweeper at its best angle with the floor. The close agreement of the two tests indicates that a carpet artificially soiled with dirt actually removed from another carpet by a vacuum cleaner is as difficult to remove as dirt which has been worked into a carpet by ordinary daily use. This condition does not result when any other substance is used to artificially soil the carpet, as will readily be seen by reference to the tests of carpets filled with sand and other substances which have been described in this chapter.

A comparative test of three different renovators was recently made by the author. Renovator No. 1 had a cleaning slot 14 in. long by ¾ in. wide, the edges of the slot being a segment of a circle having a ⅛-in. radius. This form of cleaning surface allows very small area of contact with the surface cleaned and permits the admission of large air volumes, about 56 cu. ft., with 2-in. vacuum. It is practically a Type F renovator, similar to that used in the tests at Hartford.

Renovator No. 2 had a cleaning slot 9½ in. long and ¼ in. wide, the face of the renovator being approximately ⅞ in. wide and practically a plain surface, a typical Type B renovator.

Renovator No. 3 had a cleaning slot 7¼ in. long and ⅛ in.

wide, the face of the renovator being 3/8 in. wide and the edges slightly rounded, a typical Type A renovator.

The carpet used was a Colonial velvet rug with 1/8-in. nap, closely woven, containing 6 sq. yds. This rug was filled with 12 oz. of dirt taken from separators of cleaning machines, from which the lint and litter had been screened. This was rubbed into the carpet until no dirt was visible on the surface, the surface being then lightly swept with a brush and weighed.

In cleaning this carpet the renovator was passed once over the entire surface at the rate of about 70 ft. per minute. This required six strokes and 50 seconds for No. 1 cleaner, nine strokes and 77 seconds for No. 2 cleaner, and 12 strokes and 100 seconds for No. 3 cleaner.

The carpet was then weighed, spread down and gone over three times, weighed, spread down and gone over four times. This operation was repeated until the carpet came within 1/2 oz. of its weight when received.

Each of the three renovators was operated with a vacuum of 2 in. at the renovator.

The results of these tests are illustrated by curves 1A, 2A and 3A in Fig. 23. This shows that to remove 95% of the dirt the renovator had to be passed over the carpet 20 times for No. 1 renovator, 15 times for No. 2 renovator and 8 times for No. 3 renovator.

Similar tests were then made with each of the renovators, with a vacuum of 4.5 in. of mercury at the renovator. The results are shown by curves 1B, 2B and 3B (Fig. 23) These show that to remove 95% of the dirt the renovator had to be passed over the carpet 11 times with No. 1 renovator, 6 1/2 times with No. 2, and 4 1/2 times with No. 3.

These tests are all on the same carpet, with the same quantity of the same dirt and with the renovators moved at the same speed in each case. The comparison of the results should give a fair indication of the efficiency of the different types of renovators at different degrees of vacuum within the renovator and, therefore, form the most conclusive proof of the statements relative to the efficiency of renovators as given in this chapter.

All cleaning tests that the author has observed indicate that the higher the vacuum within the renovator the more rapid and effective the cleaning, and that the efficiency of the renovator is fully as high with a small as with a large volume of air passing through the renovator and with the same degree of vacuum within same. Therefore, the most effective and eco-

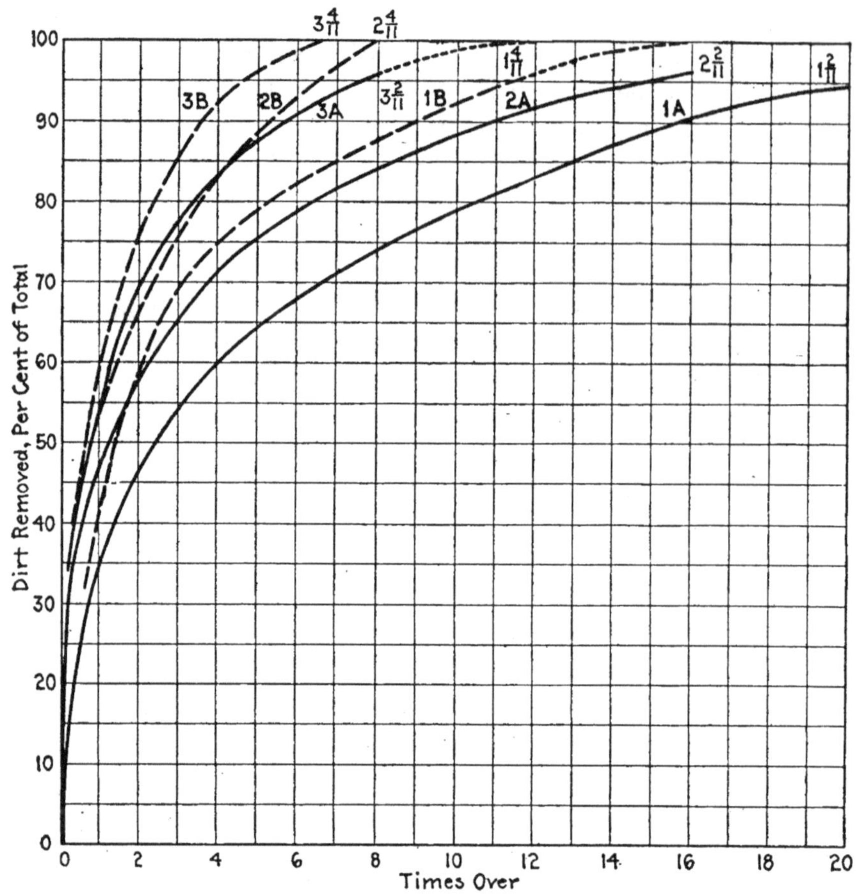

FIG. 23. TESTS SHOWING EFFICIENCY OF DIFFERENT TYPES OF RENOVATORS AT DIFFERENT DEGREES OF VACUUM.

nomical renovator should be that which gives the highest vacuum with the least air passing.

If the degree of vacuum within the renovator be carried to an abnormally high degree, there will be a tendency for the renovator to cling so close to the carpet that its operation will be difficult and the wear on the carpet rapid. The production of this high vacuum, with a larger quantity of air exhausted, will result in the expenditure of power at the renovator in excess of the gain in efficiency and speed of cleaning.

It is evident that the wider the cleaning slot, the greater will be the tendency of the renovator to stick to the carpet with a high vacuum within the same. The author has experienced no difficulty in operating the 10-in. renovator, with 3/16-in. cleaning slot, with a vacuum as high as 9 in. of mercury, but wider-slot renovators always push hard when any high degree of vacuum exists within them.

Effort Necessary to Operate Various Types of Renovators. —The author made a series of tests to determine the effort necessary to operate the various types of renovators under different conditions. In making these tests the renovator was attached to a spring balance and pulled along the floor, the pull required to move the renovator being observed by the reading of the balance. Three types of renovators were used in this test: Type A, having a cleaning slot 5/16 in. wide and 12 in. long; Type C, having a cleaning slot 5/16 in. wide and 12 in. long, with an auxiliary inrush slot ¼ in. wide and 12 in. long; Type F, having a cleaning slot ¾ in. wide and 10 in. long. The results were as follows:

TABLE 5.

EFFORT NECESSARY TO OPERATE CLEANING TOOLS.

Kind of Carpet.	Type of Renovator.	Vacuum at Renovator, In. Hg.	Pull, Pounds.	Air, cu. ft. per min.
Brussels, short........	A	8	20	27
Napped, close back...	C	6½	17	31
	F	3½	11	59
Axminster, long nap..	F	3½	14	59
Velvet, with glue.....	A	8½	18	28
Sized back...........	C	6½	17	31
Velvet, without glue..	A	3½	15	40
Sized back............	C	1	12	45
Linoleum	A	13	23	12½
	C	1	10	40

It may be noted that, when operating on the Brussels and the glue-sized velvet, the pull required to move all types of renovators bears a direct ratio to the degree of vacuum under the renovator, and that the quantity of air exhausted is the same for each renovator on either carpet, but different for each type of renovator. It is evident that, in this case, very little air enters the renovator by passing up through the carpet, and hence the action of the inrush slot on Type C renovator is noticeable only to a slight degree. When operating on velvet carpet, without glue-sized back, the inrush slot, in conjunction with the greater quantity of air coming through the carpet, has caused the passage of a large quantity of air, while the vacuum maintained at the renovator is greatly reduced over that which was maintained under Type A renovator when the same quantity of air was passing. In this case, nearly all of the air entering Type A renovator came from the under side of the carpet. The effect on the efficiency of cleaning with Type C renovator under these conditions can readily be imagined, by reference to former tests, as being greatly reduced over that of Type A when passing the same quantity of air. With linoleum, the action of the inrush slot of the Type C renovator has again greatly reduced the vacuum under the renovator, although the quantity of air is much in excess of that passing Type A renovator. The difference in the behavior of the renovators on different makes of carpet is seen to be due largely to the difference in the quantity of air which passes up through the carpet into the renovator.

It is evident that, with the same degree of vacuum within the renovator, all types are equally easy to push and that, if the vacuum within the renovator becomes higher than is necessary to produce good cleaning results, unnecessary effort will be required to operate the renovator.

Relative Damage to Carpets with Various Types of Renovators.—A few tests have been made by the author to determine the relative damage to carpets with the various types of renovators in use and it is found that, when the edges of the renovators are made exceedingly sharp, considerable nap is pulled out. However, if the edges are made slightly rounding and not too narrow, no undue wear will occur with any of the types of renovators described, provided the vacuum in the renovator is not permitted to become greater than 5 in. of mercury.

The author considers that for best results the vacuum should not be less than 3½ in. of mercury at the renovator and that at least 2 in. is necessary to do even fair work, while, to permit easy operation and prevent undue wear on the carpets, it should not be higher than 5 in.

Before deciding which type of renovator will be most economical to use in any case the character of the cleaning to be done must be considered.

Of the various types of renovators considered in this chapter, Type C can be dismissed at once, as it is neither as effective a dust remover as Types A or F nor will it remove litter any more effectively than Type F. Tests of Type D renovator do not show as good results as a dust remover as Type A, nor will it remove litter any more effectively. Type E renovator is a modification of Type C and is not likely to be any better.

The selection, therefore, lies between Type A and Type F renovators, the former being by far the best dust remover, while the latter will pick up a limited amount of small litter, such as matches, cigar and cigarette stumps, and small bits of paper. Where large quantities of these articles are likely to be encountered, it is more important that the renovator should be capable of picking them up, but, unfortunately, when these articles are met with, there are also likely to be much larger articles present that cannot be picked up by any but a specially-designed renovator, and other means must be employed to remove them.

In residences, private offices and nearly every place where carpets or rugs are likely to be used, waste baskets and cuspidors are provided and the articles mentioned are deposited in them rather than on the floor. Thus, the renovator will be required to remove dust, cigar ashes and sand or mud only, all of which can be readily removed with a Type A renovator with less expenditure of power than with a Type F renovator.

Public places, such as ante-rooms, reception rooms and other offices to which the general public is admitted in great numbers and which are sometimes carpeted, are likely to contain articles which can be picked up by Type F renovator and not by Type A. For cleaning such places, a Type F renovator is necessary, although it requires considerably more power, but the author sees no reason why this type of renovator should be used to the exclusion of Type A, even in buildings containing rooms of this character. If the building also contains several rooms where litter will not be encountered, the author would recommend that both types of renovators be used, each in its proper place, and thereby cause a considerable saving of power in cleaning rooms where no litter is encountered.

For residence work there is little need of providing carpet renovators capable of picking up litter and, also, there will be very little bare floor cleaning to be done, which requires larger volumes of air. A smaller capacity exhausting plant, therefore, can be installed, if the Type A renovator is adopted.

In large office buildings where all cleaning is done after office hours, where the building is provided with its own power plant, and where speed of cleaning and ability to clean all apartments with the fewest tools to be carried by the cleaners is desired, it appears to be better to use only Type F renovators for all carpet work, as the extra power required will not be of vital importance.

Summing up the matter, the author believes that both Type A and F renovators have their uses in their proper places but that Type A has the widest field of usefulness, yet it need not invade the field of the other. He also believes that this fact will be realized by manufacturers in the near future, when the two types of renovators will work together side by side for the general good of the manufacturers and the users.

CHAPTER IV.

Other Renovators.

The renovator which is next in importance to the carpet renovator is that used for cleaning bare floors. The earliest form of this renovator was the oscillating floor type introduced by Mr. Kenney. This was a modification of the narrow-slot carpet renovator introduced by him. The body of same was curved and supported on two small wheels or rollers, with the intention of bringing the cleaning slot close to the surface cleaned without its touching same, as indicated in Fig. 24.

This form of renovator was found to be impracticable for the reason that any change in the angle with which the stem or tube connecting the body of the renovator with the handle in relation to the surface cleaned tended to make its action

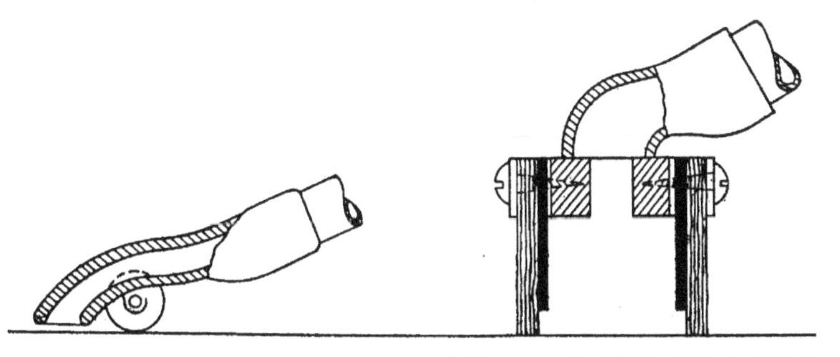

FIG. 24. EARLY TYPE OF BARE FLOOR RENOVATOR.

FIG. 25. LATER TYPE OF BARE FLOOR RENOVATOR.

ineffective. If the angle were made less the distance between the cleaning slot and the floor was increased, allowing the air to enter the cleaning slot without coming in contact with the surface to be cleaned, or, if the angle were made greater, it would cause the face of the renovator to strike and damage the surface of the floor.

The wheels or rollers on which this renovator was mounted,

being so small, were subject to rapid wear both on their faces and in their bearings, and when these wheels were slightly worn the renovator was practically useless. On account of the above defects this form of renovator was abandoned shortly after its introduction.

The next form of renovator to be tried was a modification of the ordinary soft bristle brush, such as had been in general use for cleaning hard wood floors. The bristles were arranged around the edges of the cleaning slot, in the body, which was shaped similar to the slot in the carpet renovator. Rubber or leather curtains or skirting, extending nearly to the ends of the bristles, was placed inside of these bristles in order to cause the air in entering the body of the renovator to come into intimate contact with the surface to be cleaned. The general form of this type is shown in Fig. 25.

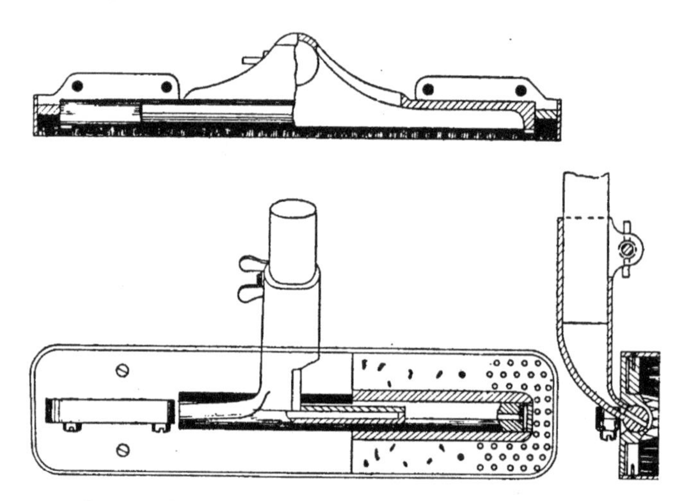

FIG. 26. ANOTHER TYPE OF BARE FLOOR RENOVATOR.

This form of renovator, while more efficient than the oscillating floor type, still had its faults in that it had a tendency to push the dirt along the floor in front of it, much the same as the floor brush from which it was copied was designed to do. Also, there was too much tendency for the air to pass into the body of the renovator without coming into intimate contact with the surface to be cleaned. While this type of floor renovator or a slight modification thereof is still in use by several manufacturers today, it never has and never will be an effective bare floor cleaner.

A modification of this type of bare floor renovator, in which the bristles have been shortened and made thicker, the skirting or flaps placed on the outside and the stem provided with a swivel joint, is shown in Fig. 26. Such an arrangement is an improvement over the former type as, owing to its wider and shorter mass of bristles, there is less tendency for the air to pass into the body of the renovator without coming into intimate contact with the surface cleaned. It is still prone to push its dirt before it and is far from being a perfect bare floor cleaner.

The next modification in the bare floor renovator was the abandoning of the bristle brush in favor of a cleaning surface composed of felt as shown in Fig. 27. In this form of reno-

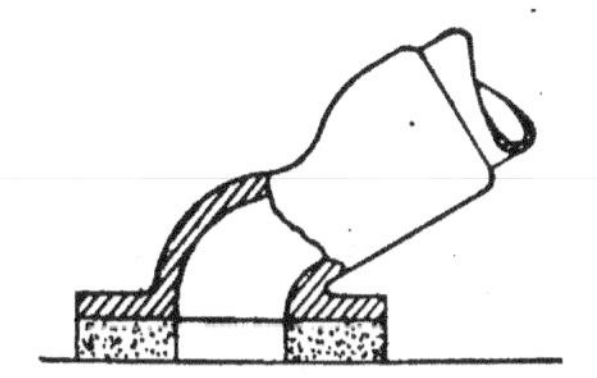

FIG. 27. BARE FLOOR RENOVATOR WITH FELT CLEANING SURFACE.

vator the air entering the body of the same must pass either between the felt and the surface cleaned or through the felt itself, and this air quantity is small. Since this renovator has a wider cleaning slot than the Type A carpet renovator, and, as it is used with the same vacuum producer, hose and pipe lines, a considerable degree of vacuum will be produced under same, especially when operated on polished floors, where the conditions are nearly the same as we observed with Type A carpet renovator operated on linoleum. With the wider slot, the effort to move these renovators becomes too great for easy operation. This trouble can be overcome by using a soft grade of felt which permits sufficient air to pass through its open pores to reduce the vacuum under same and permit easy operation. Unfortunately, this felt is subject to rapid wear when operated on surfaces as hard as floors and its use has been abandoned in favor of a harder felt. Openings are left in the felt to permit the passage of sufficient air to reduce the vacuum in the renovator to working limits. These slots have taken many forms. In one form the felt was placed in alternate X and diamond shapes, glued to the face with small open spaces between them, as illustrated in Fig. 28. However, as these

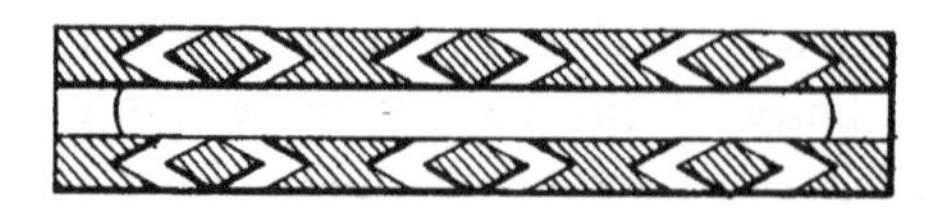

FIG. 28. BARE FLOOR RENOVATOR WITH UNUSUAL FORM OF SLOT.

small pieces must be held in place by glue, they are easily broken loose and the efficiency of the renovator impaired.

Another method, which has now become standard, is to open the ends of the renovator sufficiently to permit easy operation. This method produces high velocities at these end openings which are very effective in cleaning close to walls and in corners, where large quantities of dust always lodge and are removed with difficulty without these open slots.

The wear on these felt faced renovators was found to be so rapid that hard felt or composition rubber strips, placed so that the wear comes on the edges of the same, have been substituted. The felt or rubber was screwed on to the outside of a metal shell and projected sufficiently below the face of the metal to permit considerable wearing off of same before the

FIG. 29. BARE FLOOR RENOVATOR WITH HARD FELT OR COMPOSITION RUBBER STRIPS.

surface of the metal came in contact with the surface cleaned. When this occurs, the felt strips can readily be replaced with new ones. The ends are left open about ½ in. to form an inrush for the entering air. Such a type is shown in Fig. 29.

This renovator, in either of the above-described forms, is a great improvement over the bristle brush in that the air passing into the body of the renovator must come into intimate contact with the surfaces cleaned, but it still has the disadvantage of tending to push the dirt before it.

A modification of the above-described renovators has been introduced, in which the wearing surface of the renovator, which is covered with felt, is rounded as shown in Fig. 30. With this form of bare floor renovator, the air passing into same is not only brought into intimate contact with the surface cleaned but the dust is also crowded under the curved surface of the renovator as the same is pushed over the floor and thus brought directly into the path of the air current.

The last named type is by far the most effective for cleaning either polished or unpolished floors. It must be provided, however, with inrush slots in order to prevent its sticking and preventing easy operation. When operated with hose pipe and

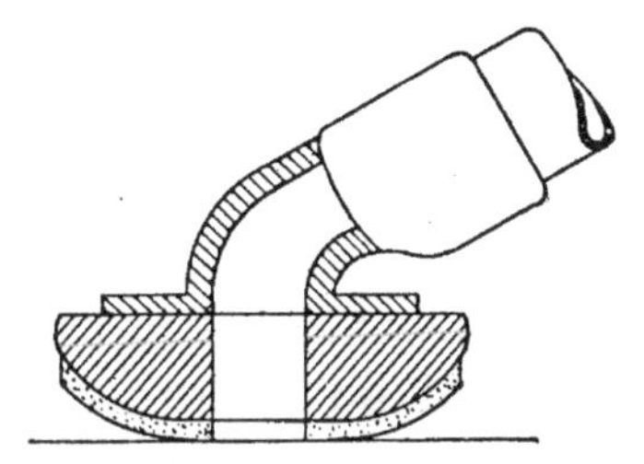

FIG. 30. BARE FLOOR RENOVATOR WITH ROUNDED WEARING SURFACE.

a vacuum producer necessary to produce 2 in. of vacuum in Type A carpet renovators, at least 30 cu. ft. of air must be permitted to pass the renovator. When operated with systems adapted to produce 4½ in. of vacuum in Type A carpet renovators, at least 70 cu. ft. of air must pass the renovator in order to permit easy operation.

This increase in the air quantity without change in the degree of vacuum in the case of these renovators, is not without increase in efficiency, as in the case of the carpet renovators, because large quantities of dust and also small litter are met with much more frequently on bare floors than on carpets. With the increase in the volume of air passing, it is possible to pick up much heavier articles than with the smaller quantity. It is also possible to pull dust out of deep cracks or from surfaces which are not in contact with the renovator face, such as the spaces between the slats of floors of trolley cars. This would not be possible with the small air quantity. The use of the larger quantity of air prohibits the use of small-sized hose and pipe and, therefore, larger articles can be conveyed through them. Where a large amount of bare floor must be rapidly cleaned the use of the larger air quantity is recommended.

A renovator (Fig. 30a) of unusual interest has recently been developed by The United Electric Company, known as the Tuec school tool. This is a bare floor tool open at both ends. It is made telescopic and is mounted on three wheels fitted with spring-actuated guide rails which are adjustable to the exact distance between the legs of school desks. A turbine motor, operated by the air passing through the renovator, is arranged to drive two of the wheels by means of worm gear and clutch.

In operation the tool is placed opposite the front of a row of desks. The clutch engaged on the turbine propels the tool through the space between the desk legs to the rear of the room. When the tool strikes the wall at the rear of the room, the clutch is disengaged and it is pulled back by drawing in the hose. The spring-actuated guides cause the cleaning slot to lengthen when passing between the desk legs thereby cleaning these spaces. The tool is then sent up the aisle, the wheels being set so that it hugs the left side of the aisle when going up and the right side when pulled back. The use of this form of tool should result in considerable saving of time in cleaning school rooms. Unfortunately, it cannot be operated where pedestal stools are used.

For use in cleaning walls, ceilings, and other flat surfaces of similar character, the bristle brush is practically the only form of renovator used.

Rubber skirting cannot be used on these brushes as it is too harsh for the easily-marred surfaces encountered by this renovator, and cotton flannel or a very soft grade of felt takes the place thereof. This change in the material used for skirting results in a greater short-circuiting of the air into the cleaner without coming into intimate contact with the surface cleaned than occurs when used with rubber or hard felt on bare floors.

As the material to be removed from surfaces of this character is very light dust, which has simply settled on the surface and is not ground in, it is very easy to dislodge. When a bristle brush, with a small volume of air passing through same, is used to remove this material, a greater portion thereof is pushed off the projections and other points of lodgment and falls to the floor from whence it must be removed by a second operation, using a floor renovator. In fact, the use of an ordinary bristle brush, followed by the use of a floor renovator, will give almost as good results as the use of a bristle wall brush with a small quantity of air passing. However, with a large quantity of air passing into the renovator, this light surface dust will all be picked up by the rapidly-moving air current and effective cleaning can be accomplished without the renovator coming into direct contact with the surface to be cleaned.

The author considers that a different form of renovator is necessary to effectively clean walls, ceilings and similar flat surfaces, with a small quantity of air passing and would recommend the use of some form of renovator having a cleaning face composed of cotton flannel or some other soft substance which could be moved over the surface cleaned, in intimate contact therewith and without damage thereto. With the soft, open fibre of the substance necessary to be used as a working surface, sufficient air would enter the renovator without resorting to the use of inrush slots or openings and much better results would be obtained. No such renovator has been designed for this purpose to date, for what reason the author does not know, and until some such renovator is produced a large volume of air will be necessary for cleaning this kind of surfaces.

An illustration of this defect in the wall brush was brought to the author's attention recently in watching a gang of laborers cleaning the walls in the U. S. Treasury Building. They had at their disposal a portable cleaner of the most efficient type, but in lieu of using the wall brush provided with same, they were rubbing off the walls with a cloth mop which had been soaked in oil, then air-dried, known as the "dustless duster." This was mounted on the end of a pole. The workmen frequently cleaned this duster with the vacuum cleaner hose without any renovator attached thereto. This cleaner, with brush in use, passed approximately 30 cu. ft. of free air per minute. It is evident that these laborers had learned by experience that it was practically useless to try to remove dust from the walls by the direct application of the wall brush to surfaces and were undoubtedly accomplishing much better results in the roundabout way they had of necessity adopted.

When carved or other relief work is encountered, the round bristle brush, with extra long bristles and cotton flannel skirting, is nearly universally used. This type of renovator is shown in Fig. 31.

FIG. 30a. THE TUEC SCHOOL TOOL.

FIG. 31. ROUND BRISTLE BRUSH FOR CARVED OR OTHER RELIEF WORK.

Owing to the irregularity of such surfaces, intimate contact therewith cannot be obtained and practically no results will be had unless there is a large quantity of air passing through the renovator. When a large quantity of air is available, nearly as good results in cleaning this character of sur-

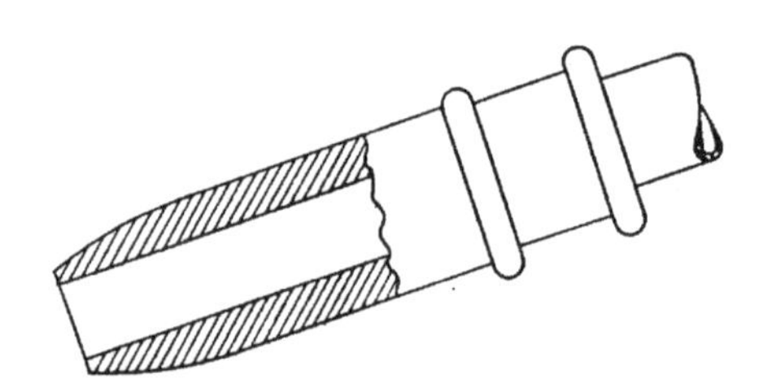

FIG. 32. RUBBER-TIPPED CORNER CLEANER FOR USE ON CARVED OR OTHER RELIEF WORK.

face can be obtained by the use of the straight rubber-tipped corner cleaner, with a round opening about ¾ in. in diameter, as illustrated in Fig. 32. A very high velocity will be obtained through this renovator which will pull the dust out of inaccessible places. This form of cleaner is also very effective for cleaning the corners of rooms, where the floor and walls intersect, veritable dust catchers that they are, the cleaning of which is fully as important as it is difficult. Pigeon holes and other small compartments in safes, desks and similar furniture can be easily cleaned with this little renovator by simply introducing it into the front of such compartment.

To be effective, this renovator must pass approximately 55 cu. ft. of air per minute and will require a vacuum within the renovator of approximately 3½ in. of mercury. Where only a small quantity of air is available, the author considers that it is better to make use of compressed air to blow the dust out of relief work, pigeon holes, and other inaccessible places and subsequently pick this dust up with other forms of renovators after it has found lodgment at more accessible points.

The cleaner which has met with the most disastrous results to the surfaces cleaned is the furniture or upholstery renovator. This has nearly always taken the form of a small carpet renovator. The type of upholstery renovator used for many years by the Sanitary Devices Manufacturing Company is illustrated in Fig. 33. This renovator had an inrush slot in the center,

FIG. 33. EARLY TYPE OF UPHOLSTERY RENOVATOR.

separated from a cleaning slot on each side by a partition extending to within 1/32 in. of the working face of the renovator. It had the hose connected into one end which was extended to form a handle. With this cleaning tool it was considered impossible to obtain a high vacuum within the renovator, as the inrush slots were supposed to act as vacuum breakers. However, as the surface of the upholstery is not firmly attached to the furniture it could be drawn up into the cleaner, closing the space under the partitions and permitting a high vacuum to be obtained. This caused the renovator to stick, but, owing to the narrow slot on each side of the inrush, the fabric was not caught.

Other manufacturers used a renovator with a single slot, in some cases as wide as ¼ in., and instances are on record where the coverings of the furniture have been drawn up through the cleaning slot into the renovator and wedged so tightly that it was necessary to cut the covering from the furniture in order to release the renovator. To overcome this difficulty one manufacturer constructed the renovator in two pieces, secured together with screws, so that, in case the renovator became caught, it could be taken apart to release the fabric.

Many manufacturers have attempted to overcome this destructive tendency of the straight-slot upholstery renovator by inserting partitions on the cleaning face of the renovator, thus dividing the cleaning slot into a number of small slots the area of each not being sufficiently large to permit the drawing in of the fabric. These cleaners have followed two general forms, one having narrow slots running lengthwise of the cleaner, as illustrated in Fig. 34. This form reduces the destructive ten-

FIG. 34. UPHOLSTERY RENOVATOR WITH NARROW SLOTS TO PREVENT DAMAGE TO FURNITURE.

dency to a great extent, but does not entirely prevent drawing the fabric into the renovator. If the partitions across the renovator be continuous, as indicated by the sketch, there will be a portion of the renovator which will not do any cleaning. Another form uses short slots, sufficiently inclined for the top of one slot to overlap the bottom of its neighbor, as shown in Fig 35. This form of renovator is effective throughout its entire length and the small area of each slot makes it practically impossible to draw the fabric into the cleaning slot. It is considered by the author to be superior to the former type, especially when cleaning lace curtains or silk hangings or any other very light fabric.

However, if the exhauster be of such characteristics and the hose and pipe lines be so proportioned that there is practically a constant vacuum in the renovator, regardless of the quantity of air passing, and provided this vacuum is not allowed to exceed 5 or 6 in. of mercury, no disastrous effects will be experienced in cleaning light-weight fabrics with a straight-slot renovator having a cleaning slot not over ¼ in. wide. The use of this type, in connection with a system having the above-described characteristics, is recommended whenever rapid cleaning is desired.

Upholstery renovators make the most serviceable clothing cleaners, while a small type of bristle brush, not over 4 in. long and not over ¾ in. wide, makes the most serviceable hat brush.

An important form of renovator is that used for cleaning between the sections and behind heating radiators. A piece of

FIG. 35. ANOTHER TYPE OF UPHOLSTERY RENOVATOR WITH SHORT SLOTS.

tubing, flattened at its outer end, is by far the most effective device for this purpose. This renovator, in connection with the hat brush tool, makes the two best renovators for use in the library, effective cleaning being possible with not more than 20 cu. ft. of air per minute, but much faster work can be done with larger quantities.

Another form of renovator sometimes furnished is the small hand brush. This is a bristle brush, approximately 8 in. long

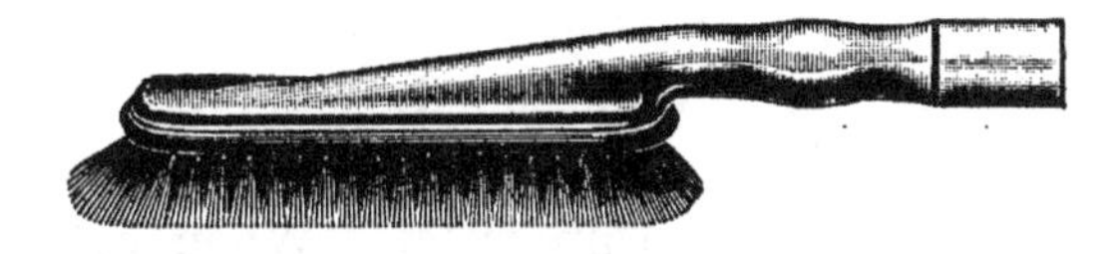

FIG. 36. HAND BRUSH TYPE OF RENOVATOR.

and 2 in. wide, with the hose connection made into one end of same, as illustrated in Fig. 36. This renovator is useful for cleaning wooden furniture, shelves, tables, and other horizontal surfaces at about hand height, but, owing to the tendency of the air to short circuit in its way to the body of the renovator, it will not do effective work with small quantities of air.

Many manufacturers have produced a special renovator for cleaning stairs. This has nearly always taken the form of a bristle brush, approximately 4 in. square. When renovators are rigidly attached to their stems, this form of renovator is convenient and almost a necessity. However, when swivel joints are provided, the ordinary carpet or bare floor renovators are fully as convenient, and, being larger, are more rapid cleaners, and the stair renovator is unnecessary.

In isolated cases, where unusual cleaning is necessary, such as the removal of cork dust from the floors of a cork factory, picking up telegraph forms from the floors of stock exchanges, picking up wrapping papers in watch factories, etc., special forms of renovators, with large openings and large capacities for air exhaustion, become necessary. These appliances have generally taken the form similar to the carpet renovator, but with much wider slots, the forward edges of which are raised slightly above the surface of the floor when the renovator is in operation. These renovators, being of no use for any other purpose than that for which they are specially designed, and requiring quantities of air in excess of those usually provided for ordinary types of renovators, may be considered simply as special appliances and do not form a part of the outfit required to be furnished with an ordinary cleaning system.

Another class of cleaning which requires a special system and special appliances is the renovation of furs. Furs must never be brushed, as it tends to mat the hair and produce an effect opposite to renovation. The only agent suitable for renovating furs is compressed air and the form of renovator best suited for this work is a straight nozzle, flattened at the end with a slot approximately 4 in. long and not over 1/32 in. wide, from which the air escapes in a thin sheet. When held at such an angle that the air will impinge on the skin under the hair, a thorough renovation of the fur is possible.

For the renovation of pillows a hollow needle, with small openings along its sides, supplied with compressed air, produces the best results. The needle is thrust through the cover into the mass of feathers, the air tending to loosen up the matted feathers and to leave them in practically the same condition as when the pillow was first filled.

As the arrangement of the air removal system, to permit it being reversed from exhaustion to compression, complicates the outfit and adds to its first cost, and as cleaning of this character is required only at rare intervals, these renovators may also be considered as special and need not be included in the average equipment.

The author considers that the renovator equipment for a system in which from 20 to 30 cu. ft. of air per minute is exhausted for each renovator in operation, and which the author classes as a "small volume" system, should contain the following renovators in each "set" furnished:

One carpet renovator with cleaning slot 1/4 in. by 12 in. long.

One bare floor renovator 12 in. long, with curved felt-covered face.

One wall renovator 12 in. long, with cotton flannel and curved face.

One upholstery renovator with slot 1/4 in. by 4 in.

One corner cleaner.

One radiator cleaner.

In addition, one or more hat brushes should be included with each installation.

The renovator equipment for a system in which 70 cu. ft. of air per minute is exhausted for each renovator in operation, which the author classes as a "large volume" system, should contain the following renovators in each "set" furnished:

One carpet renovator, with slot 1/4 in. by 15 in.

One bare floor renovator 15 in. long, with curved felt-covered face.

One wall brush, with skirted bristles 12 in. long and 2 in. wide.

One hand brush, with hose connection at end, 8 in. long and 2 in. wide.

One 4-in. round brush for relief work.

One upholstery renovator.

One corner cleaner.

One radiator tool.

At least one hat brush with each system.

The number of sets of renovators to be furnished should naturally be at least equal to the number of sweepers which the plant will handle, and in all buildings, except residences, there should be one set of renovators for each floor of the building. This will be ample, except in exceedingly large buildings.

The wearing face of any renovator should never be made of soft metal, such as brass or aluminum, as the action of the dust passing the face of the renovator, where the velocity is always the highest in the system, will roughen these parts and cause undue wear on the surfaces cleaned. Stamped steel is undoubtedly the best material for wearing surface and cast-iron ranks next. These are the only materials which should be permitted.

CHAPTER V.

Stems and Handles.

Having discussed the various forms of renovators in detail, the next appliance to be taken up is the connection between the renovator and the cleaning hose, this being the next portion of the apparatus forming a conduit for the dust-laden air on its way from the renovator to the atmosphere on the exhaust side of the vacuum producer.

In order that the renovator may be moved about on the surfaces to be cleaned, a rigid handle must be provided and, in order that these various surfaces may be reached while the operator remains in a standing position, it is necessary that this handle be of considerable, as well as variable, length. Also, a passage for the dust-laden air must be provided in connection with this handle. These conditions are best met by a metal tube, which the author terms the stem.

These stems have been made of various metals, that first used being drawn brass, probably because it is best suited to be nickel plated. On the earlier systems they were almost invariably made of No. 16-gauge tubing, ⅞-in. outside diameter, and were bent at their upper end through an angle of nearly 135° in order that the hose would hang from the stem vertically downward, when the stem was held at an angle with the floor of 45°.

The lower ends of these stems were rigidly attached to the renovator in such a manner as to assume the above-mentioned angle with the floor when the renovator was in the proper position for cleaning. In order to bring the curved portion of the stem hand high, the stem was made approximately 5 ft. long.

When operated with Type A carpet renovators, these curved stems were apparently satisfactory. However, when they were used in department stores, and other places where much bare floor cleaning was necessary, the stems were cut through at the curved portion by the sand blast action of the dust. The cutting of these stems in bare floor work, while they were satisfactory in carpet cleaning, indicates that the velocity in the stem, due to the large volume of air passing the bare floor renovator, was too great for this soft metal to withstand the impact of the dust on the curved surface. With the systems in use at that time no means were provided to control the vacuum at the vacuum producer and the hose and pipe lines were small, both of which tended to cause a wide variation in the volume of air exhausted under various conditions, in the character of surface cleaned, and in the number of renovators in use. Therefore, the value of this destructive velocity is not readily obtainable. However, the author considers that, in extreme cases, the quantity of air passing through these stems may have been as high as 55 cu. ft. per minute. As the inside diameter of the stems was ¼ in. the area was 0.44 sq. in., or 0.00328 sq. ft., and the velocity through the stem was nearly 17,000 ft. per minute. With an average air passage of 40 cu. ft. per minute the velocity was 12,200 ft. per minute.

Referring to tests of carpet renovators, Chapter III, it will be noted that the maximum volume of air passing through carpet renovators of Type A was 33 cu. ft. per minute, which gives a velocity of 10,000 ft. per minute. Apparently, at this velocity, the cutting action, due to the impact of the dust on the curved surfaces, was not severe. However, the author considers that the maximum velocity that should be permitted through these stems is 9,000 ft. per minute.

As the dirt picked up must be lifted almost vertically, the velocity in the stem must not become too low or dirt will lodge in the stem. Experiments made by the author indicate that the minimum velocity should be at least 4,000 ft. per minute, in order to insure a clean stem at all times.

Shortly after the introduction of vacuum cleaning, the use of drawn-steel tubing for the manufacture of stems for cleaning tools was standard with one manufacturer and, lately, its use has become almost universal, except in cases where very long stems are necessary, as on wall brushes when cleaning very high ceilings. For such work, aluminum stems have been adopted.

This harder metal will better withstand the cutting action of the dust and can also be made much thinner and lighter in weight than brass tubing of equal strength. These stems were made from 1 in. outside diameter, No. 21 gauge tubing, having an internal area of 0.68 sq. in., and the author does not know of any cases where these stems have been cut by the impact of the dust.

Stems of this metal are recommended by the author for use with all floor renovators and with wall brushes, except in cases where exceedingly long stems are required, when those of drawn aluminum tubing are recommended.

For use with Type A renovators, where the minimum air quantity is approximately 22 cu. ft. per minute, the greatest area permissible is $\frac{22}{4000} = 0.0055$ sq. ft., or 0.79 sq. in., equivalent to 1-in. diameter. With a maximum air quantity, under proper control, of 39 cu. ft. per minute, the minimum area will be $\frac{39}{9000} = 0.00433$ sq. ft. or 0.625 sq. in., equivalent to 0.89 in. diameter, so that a 1-in. outside diameter stem of No. 21 gauge metal, having an inside diameter of 0.932 in., is recommended.

For use with a Type F renovator, with a minimum air quantity of 44 cu. ft. per minute, the maximum area of the stem will be $\frac{44}{4000} = 0.011$ sq. ft., or 1.58 sq. in., equivalent to 1.4 in. diameter, while, with a maximum air quantity of 70 cu. ft. per minute, the minimum area will be $\frac{70}{9000} = 0.0077$ sq. ft., or 1.11 sq. in., equivalent to 1.18 in. diameter, and a 1¼-in. diameter stem of No. 21 gauge metal, having an inside diameter of 1.18 in. is recommended.

Tests of Mr. S. A. Reeve, which are discussed in Chapter III, indicate that both edges of the cleaning slot on any renovator must be in contact with the surface cleaned in order to do effective cleaning. A renovator which is rigidly connected to its stem can be effectively operated with the stem at but one angle with the surface cleaned, which makes the cleaning under furniture, or on wall at various heights above the floor, impossible. In order to do effective cleaning with any degree of speed and comfort to the operator, some form of swivel joint between the renovator and its stem is necessary.

These swivels have been made in many forms, one of which consists of two hemispheres connected by a bolt on their axis, as shown in Fig. 37. This form of swivel is unsuited for use under these conditions, as lint, thread and any other small articles picked up will catch on the bolt which lies directly in the path of the dust-laden air current, and its use should be prohibited in all cases.

Another form of swivel, which is must better than the last mentioned, is shown in the illustration of the bare floor brush, Fig. 26, Chapter IV, there being no obstruction in the air passage. However, these swivels are composed of moving parts

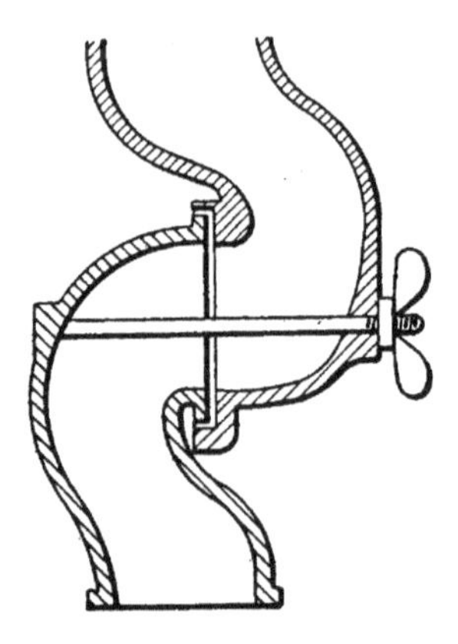

FIG. 37. FORM OF SWIVEL JOINT CONNECTING STEM TO RENOVATOR.

which are in contact with the dust-laden air and great care must be taken in their design so that in action dust does not lodge between the wearing surfaces and shortly ruin the swivel. This can be guarded against by making any opening between the parts of the swivel point away from the dust current, as indicated in Fig. 38, in which the direction of the air current is indicated by the arrow. A slightly loose fit between the wearing surfaces will permit a small leakage of air through the joint which will tend to remove any dust which may find its way into the joint. However, it is not considered advisable either to allow very much leakage through the joint, as it reduces the net efficiency of the system, or to depend much on the air current through the joint keeping the wearing surfaces clean. The swivel indicated in the illustration of the floor brush does not entirely prevent the dust entering same and it permits the movement of the stem in a vertical plane only. On the other hand, a swivel consisting of a 45° elbow, rigidly attached to the stem and turning freely on a horizontal spud, and fastened to the renovator, as shown in Fig. 38, allows a motion of the stem either in a vertical plane, which will cause the renovator to rotate, and enable the operator to pass same around or back of legs of furniture, or a semi-rotary motion may be imparted to the stem, which will permit the renovator

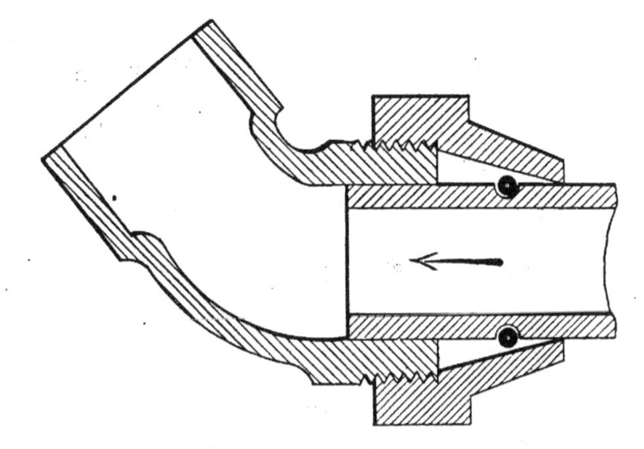

FIG. 38. SWIVEL JOINT ARRANGED TO PREVENT DUST LODGING BETWEEN THE WEARING SURFACES.

to move forward in a straight line while the angle which the stem makes with the floor will constantly decrease. After a little practice the operator can place a renovator equipped with one of these swivels in almost any position without inconvenience. Illustrations of the possibilities of this form of swivel are presented in Figs. 39 and 40, in which an operator is shown cleaning the treads and risers of a stairway without changing her position, and in Fig. 41, where the operator is cleaning the trim of a door with apparent ease. The author considers that this form of swivel is the only satisfactory joint between the renovator and its stem. It is being rapidly adopted by nearly every manufacturer of vacuum cleaners.

In operating any renovator it is nearly always drawn backwards and forwards in front of the operator, across the surface to be cleaned. When the hose is rigidly attached to the upper end of the stem, it becomes necessary to drag at least a portion of the cleaning hose along with the renovator when it is moved forward, and to crowd the same back on itself when the renovator is moved backward. This action has a tendency

FIG. 39. SWIVEL JOINT IN USE.

to kink or snarl the hose about itself and makes the operation of the renovator very awkward, often causing the operator's feet to become entangled in the hose.

This action also brings an undue amount of wear on the hose near the end which is attached to the stem, as may be readily noted by inspection of hose used with rigidly-attached stems. This will show that the end of the hose is entirely worn through, while the remainder of the hose is still in serviceable condition.

The trouble above stated can be overcome by providing a swivel joint at the point of connection between the hose and the stem. A few attempts to use a joint similar to that first described in connection with the renovator and its stem, as illustrated in Fig. 37, have been made, but without much success, as the bolt through the air passage catches dirt and there is not sufficient freedom of movement between the portions of the swivel. Variations of this form of joint have been made, one of which is provided with a screwed union to join the two

FIG. 40. ANOTHER USE OF SWIVEL JOINT, SHOWING POSSIBILITIES OF THIS FORM.

portions, as shown in Fig. 42. This is a much better form than that first described and has been successfully used in connection with heavy 1-in. diameter hose. Care must be exercised that the direction of the flow of air is always in the direction indicated by the arrows in the sketches, as a reversal, if only for a short time, will ruin the joint, due to lodgment of dust in the moving parts.

Still another variation in this form of swivel has the two main parts made to fit one within the other and a snap ring is placed in a groove in the male portion of the joint, this groove being deep enough to take the entire thickness of the ring. The two parts are then fitted together and the ring snaps out into a corresponding groove in the female portion of the joint, uniting the two parts. This joint gives a fairly free movement

FIG. 41. OPERATOR CLEANING TRIM OF DOOR WITH SWIVEL JOINT.

to the parts thereof, but has the disadvantage that it cannot be taken apart without breaking one of its parts.

A modification of this form of swivel has been made by the manufacturers of the last-described swivel, in which semicircular grooves have been cut, one on the inside of the female

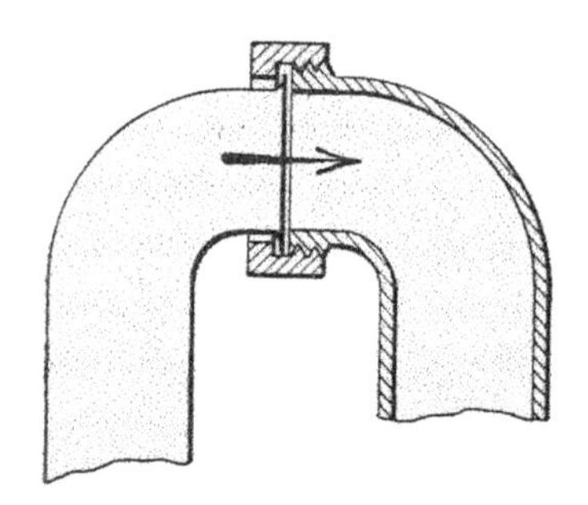

FIG. 42. SWIVEL JOINT, WITH SCREWED UNION.

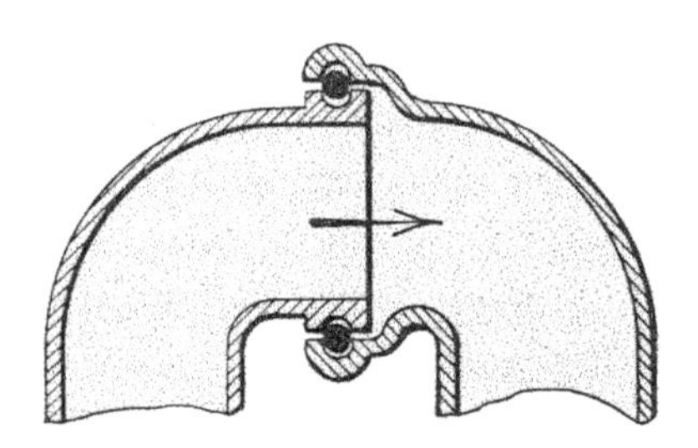

FIG. 43. SWIVEL JOINT HAVING BALL BEARINGS.

portion and one on the outside of the male portion. Steel balls are forced into this groove, after the parts are assembled, through an opening provided in the edges of the parts. This opening is closed, after the balls are in place, by a small pin, as shown in Fig. 43. The swivel then becomes a ball-bearing

FIG. 44. ACTON OF BALL-BEARING SWIVEL JOINT.

joint, with a freedom of motion characteristic of such bearings. This joint readily responds to every movement of the stem and keeps the hose hanging vertically downward and always free from kinks. Its action is illustrated in Fig. 44, in which it is being used in connection with a carpet renovator. This joint is considered to be the most efficient on the market. It is protected by a patent controlled by a manufacturer of vacuum cleaners.

Valves are placed at the upper end of the stems by many manufacturers, to cut off the suction when carrying the renovators from room to room, and when it is necessary to stop sweeping to move furniture. These valves have nearly always taken the form of a plug cock with tee or knurled handle. They are useful on large installations, where vacuum control is either inherent in the exhauster or where some means of vacuum control is provided, as a considerable saving of power may be obtained by closing same, as will be explained in a later chapter, and to overcome the unpleasant hissing noise caused by the inrush of air into the renovator when same is held off the floor.

When the exhauster has a capacity of but one sweeper and when the cleaning is done at times when the building is unoccupied, there seems to be little need for this refinement, which has two defects: first, the operators will not close the valves; second, when they have been closed they are only partly opened, as indicated in Fig. 45. When this occurs, the portions of the

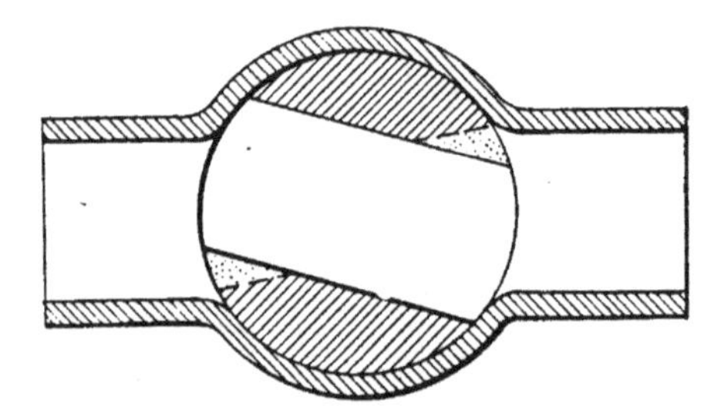

FIG. 45. ILLUSTRATION OF DEFECTS OF PLUG COCKS.

plug, which are shown stippled, are quickly cut away by the sand-blast action of the dust, making it necessary to open the valve a still smaller amount the next time it is operated, cutting off still more of the plug until a new plug is necessary in order to make the valve again operative.

A few attempts have been made to overcome these defects by making the valves self-closing and having them so constructed that when the operator grasps the handle the valve will be forced wide open, on the principle of the pistol grip. These valves will, of course, close whenever the handle is released, and it is impossible to grasp the handle in any degree

hout throwing the valve wide open. However,
is closed by a spring, considerable pressure must
he handle in order to keep it open and it acts
Sandow dumb bell in producing fatigue of the
ort time; they have not come into general use.
es in the renovator handle is considered by the
n expense not justified by the gain in economy
no longer included in specifications prepared

CHAPTER VI.

HOSE.

The more important steps in the evolution of the modern vacuum cleaning system can each be attributed to a change in the design or construction of some one of its component parts, which, in their former standard design, have acted as a limiting factor governing the form and size of other and more important parts of the system.

That part of the early systems which played the most important role as a limiting factor was one for whose production the builder of the system had to look to other manufacturers: namely, the flexible hose connecting the renovator stem to the rigid pipe lines and vacuum producer.

The early builders of vacuum cleaning systems naturally adopted a standard article for use as a flexible conduit; that is, the vacuum hose which had been used as suction lines for pumps of various characters. For such use it was not necessary that the hose be moved about to any great extent and, therefore, its weight was not an important factor and had been sacrificed to strength to withstand collapse and the rough handling to which suction hose is subject.

This standard hose was built up of many layers of canvas wound around a rubber tube or lining. A spiral wire was imbedded between the layers of canvas to prevent collapse and the whole was provided with an outer covering of rubber. Generally five to seven layers of canvas were used and the resulting hose was not highly flexible.

When used as a flexible conduit in connection with a vacuum cleaning system it became necessary to constantly move the hose back and forth and around the room to be cleaned. It was also necessary to limit the weight of the hose to that which could be easily handled by one person. This led to the adoption of small sizes of the then standard hose, ¾-in. diameter being first used, but soon this was abandoned in favor of 1-in. diameter hose weighing nearly 1 lb. per foot of length, which is the maximum weight that can be conveniently handled by one person. This size hose has become the standard for all systems maintaining a vacuum at the separators of 10. in of mercury or more.

Owing to its lack of flexibility this type of hose is easily kinked and is damaged by the pulling out of such kinks, causing the tubing or lining to become separated from the canvas and to collapse, rendering the hose useless. There is also considerable wear at the point of connection to the stems of renovators, where rigid connections are used.

The outside of this hose, being rubber, is always liberally covered with soap-stone when it leaves the manufacturer, and when new hose is dragged about over carpets, it frequently soils same to a greater degree than they are cleaned by the renovator. When this hose has been in use about twice as long as is necessary to wear off the soap-stone, its appearance becomes far from handsome and is not considered to be in keeping with the nickel-plated appliances which are furnished with the cleaning tools. To overcome this objection, an outer braid has been applied generally over the rubber coating, thus adding further to its already great weight.

What was perhaps the first type of hose to be produced especially for use with vacuum cleaning systems was that in which the fabric was woven in layers, instead of being wrapped spirally around the central tube or lining. Steam was introduced into the lining, vulcanizing the lining and firmly uniting the whole mass. This hose was made 1 in. in diameter, without any metal re-inforcement, and was covered with the usual rubber coating and with braid, when ordered. This hose weighed 12 oz. per lineal foot and 1-in. diameter was still the largest that could be easily handled.

The first attempt to produce a light-weight hose for use with vacuum cleaning systems was by covering a spiral steel tape with canvas. The air leakage through this hose was found to be so high that its use resulted in loss of efficiency of the cleaning plant and it was found necessary to line the hose with rubber. This rubber-lined hose is made in larger sizes than formerly used and 2-in. diameter hose weighs approximately 14 oz. per lineal foot. It is also much more flexible than the 1-in. hose formerly used.

The introduction of this type made it possible to use larger hose in connection with vacuum cleaning systems and permitted the use of a lower vacuum at the separators, with the same results at the carpet renovator, and a larger quantity of air when using the brushes and other renovators. Without this type of hose the low-vacuum, large-volume systems would be impractical.

Another type of hose has been recently introduced in which a wire is woven into the fabric of the hose and the rubber lining vulcanized into place as already described. No outer coating of rubber is used and, therefore, no braid is necessary. This gives a light-weight hose of great flexibility and neat appearance and is undoubtedly the best hose for residence work. It is more costly than the steel tape hose which is recommended for office building and factory use, where appearance is not important.

Hose Couplings.—The earlier systems used couplings having screw-threaded ground joints, similar to those which were then in use on hose intended to withstand pressure. These couplings

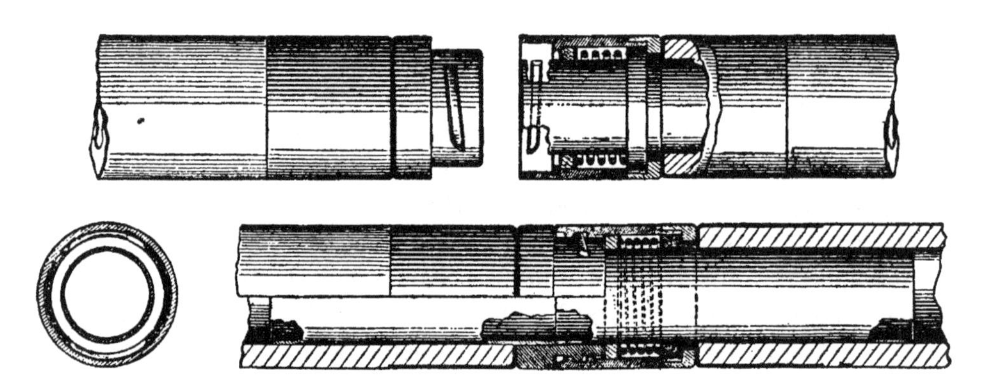

FIG. 46. BAYONET TYPE OF HOSE COUPLING, INTRODUCED BY THE AMERICAN AIR CLEANING COMPANY.

require considerable time to connect and disconnect and the threads are easily damaged by dragging the hose about. The exposed metal parts of the couplings are liable to scratch furniture.

To overcome the time required to connect and disconnect the screw-coupling, the American Air Cleaning Company introduced the bayonet type of coupling, as illustrated in Fig. 46. This coupling is not readily damaged by rough handling, but it has metal surfaces exposed which will scratch furniture.

Both of these couplings have the disadvantage that the air current in the hose must always be in the same direction and the same end of the hose must always be next to the renovator handle. Both of these features tend to increase the wear on the hose, and the reversal of the air current to remove stoppages is not possible.

The coupling produced by the Sanitary Devices Manufacturing Company has a piece of steel tubing fitted into each end of the hose and secured by means of a brass slip-coupler fitting over the tubing. All ends being alike, the reversal of the hose is possible with this form of coupling. However, the metal coupler is liable to mar furniture and sometimes there is trouble with the couplings pulling apart.

Much of the hose in use today is provided with "pure gum" ends are vulcanized in place, it is necessary to take the hose of metal tubing is slipped inside of these ends to make a coupling. With this arrangement there is no metal exposed to mar

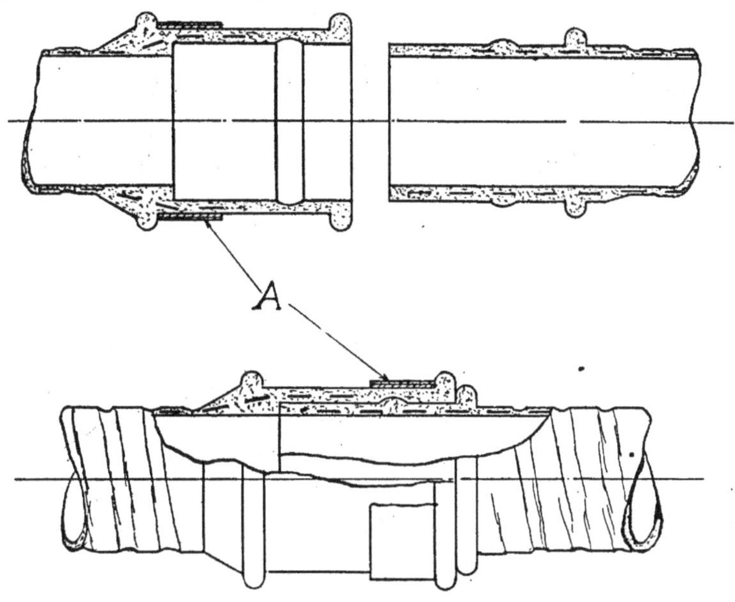

FIG. 47. ALL RUBBER HOSE COUPLING USED BY THE SPENCER TURBINE CLEANER COMPANY.

furniture and the hose lengths are reversible. However, there is some trouble from the couplings pulling apart. Since these ends are vulcanized in place, it is necessary to take the hose to a rubber repair shop whenever the hose breaks back of the coupling, which occurs frequently when rigidly attached to the stem of the renovator. These repair shops are much more numerous than a few years ago and this drawback is not a serious one.

Another form of coupling used by the Spencer Turbine Cleaner Company is the all-rubber male and female end, as illustrated in Fig. 47. This has the advantage over the metal-slip couplings and the coupling with pure gum ends in that when it is properly locked it cannot be pulled apart. It is absolutely air tight, which is true of no other coupling. But it does not permit the reversal of the hose and is, therefore, recommended for use only with hose of 1¼-in. diameter or larger, where there is less liability of stoppage, and where the ball-bearing swivel is used at the connection to the stem, preventing excessive wear at this point. The pure gum ends, with the internal-slip coupler, is considered to be the most satisfactory for use in all cases, except as above stated.

Hose Friction.—Hose friction plays an important part in the action of any vacuum cleaning system. In fact, where 1-in. hose is used, it becomes a limiting factor in the capacity of the system to perform some kinds of cleaning.

There are several tables of hose friction published by the manufacturers of vacuum cleaning systems, all of which appear to have been based on a constant velocity within the hose equal to that which would be obtained if the air were at atmospheric pressure throughout the entire length of the hose. But in practice the air is admitted to the hose from the renovator at a considerably lower absolute pressure of from 25 in. to 27 in. of mercury, and is, therefore, moving at a higher velocity. As the pressure is decreased by the friction loss in the hose, the velocity constantly increases with the expansion of the air.

The results of many tests made by the author during the past seven years, with hose ranging from 1-in. to 2-in. diameter and with an entering vacuum ranging from 0 to 7 in. of mercury and a friction loss of from 1 in. to 25 in. of mercury, indicate a close agreement with the formula given in Prof. William Kent's "Mechanical Engineer's Pocketbook," which is based on the formula:

$$Q=c\sqrt{\frac{pd^5}{wL}}$$

Q = free air in cubic feet per minute.

c = a constant which was determined by D'Arcy as approximately 60.

p=the loss of pressure in pounds per square inch.

d = the diameter of pipe in inches.

L = the length of pipe in feet.

w = the density of the entering air in pounds per cubic foot.

Reducing the pressure loss to inches of mercury and using in lieu of w, r which is the ratio of the average absolute pressure in the pipe to atmospheric pressure, this formula becomes:

$$Q=310.3\sqrt{\frac{pd^5}{Lr}}$$

To permit the rapid calculation of the air quantity which can be passed through a hose, the author has prepared the diagram shown in Fig. 48. To use this table, look up the friction loss in the hose in the right hand margin, pass along the horizontal line to the left until it intersects the line inclined at an angle of 45° toward the left, indicating the length of the hose. From this intersection pass vertically to the line inclined at approximately 30° toward the left, representing the diameter of the hose. The quantity in the left-hand margin, opposite the horizontal passing through this intersection, represents the quantity of air which would pass through this hose in cubic feet at the average density in the hose. To correct this quantity to free air, step off the distance on the vertical line from the bottom of the table, representing the average degree of vacuum in the hose, to its intersection with the curved line near the bottom of table. Transfer this distance vertically downward on the left hand margin from the quantity first read on this margin. The quantity opposite the lower end of this distance will be the cubic feet of free air per minute passing through the hose under these conditions.

The line inclined towards the right, which passes through the intersection of lines representing hose diameter, and the horizontal line representing the cubic feet of air passing through the hose at actual density in same, shows the actual velocity in the hose in feet per second.

For friction loss over 10 in. of mercury, use the figures at the right hand of the lower margin, instead of those in the right hand margin, and pass vertically to the hose diameter.

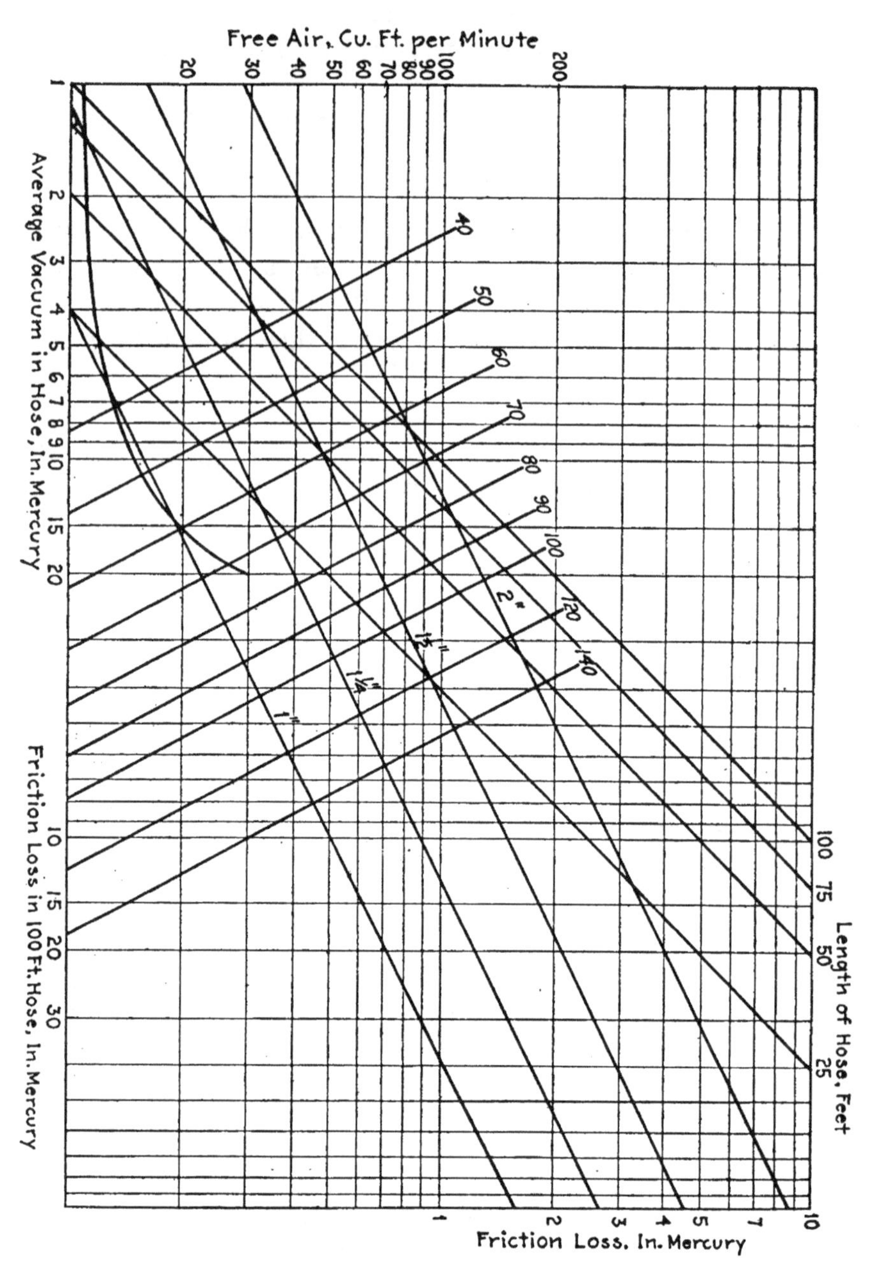

FIG. 48. CHART FOR DETERMINING HOSE FRICTION.

Then proceed as before. As these high frictions are seldom used in practice, this departure has been made in order to reduce the size of the diagram.

To illustrate how much the friction tables, based on air at atmospheric density, vary from actual results, two tests made by the author are given. In the first test it was desired to pass 68 cu. ft. of free air per minute through a 7/8-in. diameter orifice at the end of 100 ft. of 1-in. diameter hose. Tests on larger hose showed that, to permit this quantity of air to pass through the orifice, a vacuum at the orifice of 2.6 in. mercury was necessary. The most rational table the writer could find indicated that the friction loss in the hose should be 18 in. mercury, and the final vacuum necessary at the hose cock would have to be 20.6 in. mercury. On test it was found that, with 24.8 in. vacuum at the hose cock, but 50 cu. ft. of free air per minute was passing, with a vacuum at the orifice of 1.6 in. mercury, showing a friction loss of 23.2 in. mercury. With the smaller quantity of air passing, the same friction table indicated a friction loss, with this quantity of air, of but 9.8 in. mercury, or 39% of that actually observed. Checking the results of the test with the diagram (Fig. 48) gives 50 cu. ft. of free air, with a friction loss of 23 in. mercury.

To illustrate more clearly the effect of the increase of velocity on the friction loss, the actual vacuum in the hose has been computed for each 10 ft. of its length and curves drawn through these points. The results are shown in Fig. 49. The straight line indicates the vacuum which should exist were the velocity in the hose constant throughout its length, and the curved line shows the vacuum in the hose when the effect of the increasing velocity, due to the rarefaction of the air, is considered. The wide variation in the results shows clearly the error in the former assumption of a constant velocity in the hose throughout its length.

Another test, in which 44 cu. ft. of free air was passed through 100 ft. of 1-in. diameter hose, is shown graphically in Fig. 50, which discloses that the assumption of a constant velocity in the hose produces an error of 35% in the results, indicating a loss of but 7.8 in., when the actual loss is 12 in. mercury.

Naturally, the lower the final vacuum at the hose cock, the less will be the error due to the assumption of constant velocity in the hose. Tests with 1½-in. hose gave results which agree substantially with the result given in tables already published, and it was this condition that led to the discovery of the error in the assumption stated.

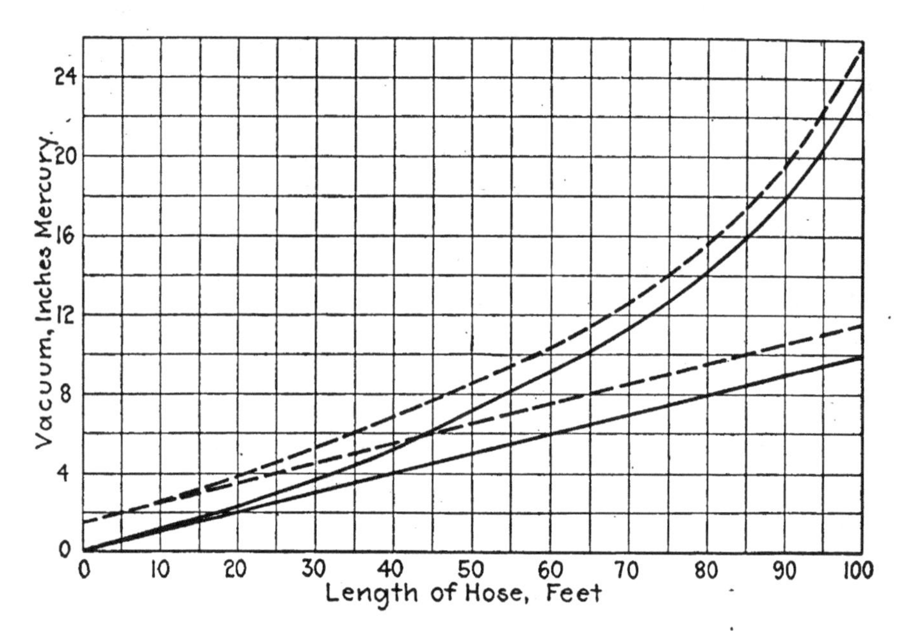

FIG. 49. EFFECT OF INCREASE OF VELOCITY ON THE FRICTION LOSS.

Effect of Hose Friction.—As any increase in the degree of vacuum necessary to be maintained at the vacuum producer over that maintained within the renovator requires a greater expenditure of power, without any increase in the efficiency or speed of cleaning, it is essential that the friction loss in the air conduit from the renovator to the vacuum producer should be made as small as possible. The friction loss in the hose is the greatest loss in any part of the system, being the smallest in diameter, and its reduction to the lowest figure possible is of vital importance.

Take, for example, the use of a Type A renovator with a vacuum within the renovator of 4½ in. mercury and with 29 cu. ft. of air passing through same. The friction loss, with varying lengths of different-sized hose, will be as follows:

TABLE 6.

VACUUM AT HOSE COCK WITH TYPE A RENOVATORS AND WITH VARYING LENGTHS OF DIFFERENT-SIZED HOSE.

Size of Hose. In. Diameter.	Length, in Feet.			
	100	75	50	25
	Vacuum at hose cock, in. hg.			
1	10	8½	7	5½
1¼	6	5.7	5.25	4.85
1½	5.0	4.85	4.75	4.62

This indicates, first, that a much lower friction loss will result with the use of larger hose than is the case with the smaller size. Note, also, that the difference in the final vacuum at the hose cock is much more uniform when the larger-sized

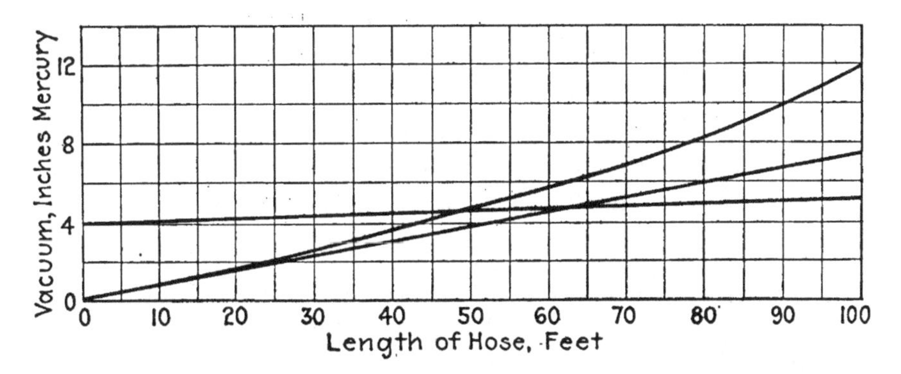

FIG. 50. ANOTHER TEST SHOWING FRICTION LOSS DUE TO VELOCITY.

hose is used in varying lengths. Since it is desired to maintain a constant vacuum at the renovator at all times and it is also desirable to be able to vary the length of hose to suit the conditions of the work, while it is not convenient to vary the vacuum at the hose cock, much more uniform results will be possible when larger hose is used. If the smaller hose is used in varying lengths and a practically uniform vacuum is maintained at the hose cock, the quantity of air and the vacuum at the renovator will vary. If 1-in. hose is used and the vacuum at the hose cock be maintained at 10 in. mercury, the air quantities and vacuum at the renovator will be approximately:

TABLE 7.

AIR QUANTITIES AND VACUUM AT RENOVATOR WITH 1-IN. HOSE AND 10-IN. VACUUM AT HOSE COCK.

Length of Hose, feet.	Vacuum at Renovator, in. hg.	Air, cu. ft.	H. P. at Hose Cock.
100	4½	29	0.80
75	5	32	0.885
50	6½	34	0.94
25	7½	37	1.02

From this it is evident that the vacuum within the renovator will be increased above that necessary for economical cleaning. It will require somewhat more effort to push the cleaner over the carpet and also a slightly greater expenditure of power at the hose cock to operate the cleaner with a short than with a long hose. However, the author does not consider that either the increase of effort to push the renovator or the increase of power will be sufficient to prohibit the use of 1-in. hose with the Type A renovator.

If we use 1¼-in. hose with Type A renovator and maintain a vacuum of 6 in. of mercury at the hose cock, the resulting vacuum and air displacement at the renovator will be:

TABLE 8.

AIR QUANTITIES AND VACUUM AT RENOVATOR WITH 1¼-IN. HOSE AND 6-IN. VACUUM AT HOSE COCK.

Length of Hose, feet.	Vacuum at Renovator, in. hg.	Air, cu. ft.	H. P. at Hose Cock.
100	4½	29	0.43
75	4.7	30	0.445
50	5.0	33	0.448
25	5.4	35	0.518

This table shows a more uniform degree of vacuum at the renovator with the varying length of hose, but the greatest difference is in the horse power required at the hose cock to accomplish the same results at the renovator.

If we use 1½-in. hose with Type A renovator, the vacuum at the hose cock can be reduced to 5 in. mercury and a practically constant vacuum will be obtained at the renovator, with an expenditure of 0.36 H. P. at the hose cock.

With the Type C renovator where the vacuum within the renovator is maintained at 4 in. mercury, with 44 cu. ft. of free air per minute passing through the renovator, the resulting vacuum at the hose cock, with various lengths of the three sizes of hose, will be as follows:

TABLE 9.

VACUUM AT HOSE COCK, WITH TYPE C RENOVATORS AND VARIOUS LENGTHS OF THREE SIZES OF HOSE.

Size of Hose, In. Diameter.	Length, in Feet.			
	100	75	50	25
	Vacuum at hose cock, in. hg.			
1	19	14	10	6.7
1¼	7.5	6.25	5.5	4.7
1½	5.1	4.80	4.50	4.25

Referring to Fig. 17, Chapter III, it will be noted that Type C renovator will not accomplish much in the way of cleaning with a vacuum in the renovator lower than 4 in. mercury. Therefore, if we use this type of renovator, with 1-in. diameter hose, its length should be limited to 50 ft., for if we use a vacuum higher than 10 in. at the hose cock, there will be too much increase in the vacuum at the renovator when short hose is used to allow easy operation, and if we use longer hose with 10-in. vacuum at the hose cock, there will be a reduction in the vacuum at the renovator and effective cleaning cannot be accomplished. Also, the power required at the hose cock to pass 44 cu. ft. of air, with a vacuum of 19 in. mercury, required to produce a vacuum of 4 in. at the renovator with 100 ft. of 1-in. hose, will be 3.3 H. P., which is prohibitive when compared with that required with the use of larger hose, i. e., 0.825 H. P. with 1¼-in. hose and 0.59 H. P. with 1½-in. hose.

The Type F renovators tested by the author will show even wider variations in the vacuum required at the hose cock with the various lengths and diameters of hose than is given for Type C renovator. However, the type F renovator, which is now used by the Spencer Turbine Cleaner Company, having a cleaning slot 15 in. long and ½ in. wide throughout its length.

passes 44 cu. ft. of free air per minute, with a vacuum under the renovator of 4 in. mercury and the resulting vacuum at the hose cock will be the same as that given in the case of the Type C renovator.

When a bare floor renovator of the bristle-brush type is attached to the hose, the effect is practically the same as when the end of the hose is left wide open, as the open character of the brush prevents the formation of any vacuum in the renovator. Therefore, sufficient air must pass through the renovator to create a friction loss in the hose equal to the vacuum at the hose cock.

As practically all systems are arranged to maintain a constant vacuum at the vacuum producer and as the pipe friction is generally less than the hose friction, the vacuum at the hose cock will be practically the same when operating a floor brush as with a carpet renovator.

Assuming that 10 in. mercury is maintained at the hose cock with 1-in hose, 6 in. with 1¼-in. hose, and 5 in. with 1½-in. hose, the quantity of air which will pass through a floor brush with various sizes and lengths of hose will be:

TABLE 10.

AIR QUANTITIES THROUGH FLOOR BRUSH OPERATED IN CONJUNCTION WITH TYPE A RENOVATORS.

Size of Hose, In. Diameter.	Hose Length, in Feet.			
	100	75	50	25
	Cubic feet of free air per minute.			
1	42	48	60	86
1¼	62	72	86	125
1½	95	110	135	190

The quantities given for the shorter hose lengths are higher than will be observed in actual practice, due to the increase in the pipe friction, which will depend on the length of the pipe lines. However, the results will illustrate the great increase in the quantity of air which will pass these bare floor brushes when operated on the same system with carpet renovators. If the same number of bare floor renovators are to be used at one time as there will be carpet renovators at some other time, that is, if the sweeper capacity must be maintained when using bare floor brushes as when using carpet renovators, a much larger air exhausting plant must be installed than would be necessary to operate that number of carpet renovators.

If it were possible to so arrange the schedule of cleaning operations that bare floor brushes were never used at the same time as carpet renovators, the vacuum at the machine might be reduced when operating the floor brushes to a point that would reduce the quantity of air passing to within the capacity of a machine designed to operate the same number of carpet renovators. Unfortunately, this condition rarely exists and, therefore, the vacuum must be maintained at the degree necessary to operate the carpet renovators that may be in use at the same time with the floor brushes.

It is also evident that if the length of hose used with bare floor brushes could be limited to the maximum ever used with the carpet renovators, a reduction in the capacity of the exhauster necessary could be made. This is another condition which the designer of the system cannot control.

Most Economical Hose Size for Carpet and Floor Renovators.—The horse power required at the hose cock to operate the bare floor brushes with each of the different sizes and lengths of hose is:

TABLE 11.

HORSE POWER REQUIRED AT HOSE COCK TO OPERATE BARE FLOOR BRUSHES IN CONJUNCTION WITH TYPE A RENOVATORS.

Size of Hose, In. Diameter.	Length, in Feet.			
	100	75	50	25
	Horse power at hose cock.			
1	1.16	1.32	1.65	2.38
1¼	0.92	1.06	1.27	1.38
1½	1.15	1.32	1.62	2.28

This shows that where bare floor or wall brushes of the bristle type are used in conjunction with carpet renovators on any system and with Type A carpet renovator, 1¼-in. diameter hose will give the lowest power consumption.

When either Type C or F renovator is used in combination with bristle-type brushes, the use of 1-in. diameter hose must be abandoned in lengths over 50 ft. and the vacuum at the hose cock must be maintained at 10 in. mercury. With 1¼-in. hose, it will be necessary to maintain a vacuum at the hose cock of 7 in. mercury, and, with 1½-in. hose, 5 in. will be sufficient, provided we continue to use 100 ft. of hose in the case of the larger sizes. The free air passing a brush type of bare floor renovator under these conditions will be:

TABLE 12.

FREE AIR PASSING BRUSH TYPE OF BARE FLOOR RENOVATOR OPERATED IN CONJUNCTION WITH TYPE C RENOVATORS.

Size of Hose, In. Diameter.	Length, in Feet.			
	100	75	50	25
	Cubic feet of free air per minute.			
1	42	48	60	86
1¼	68	76	92	130
1½	95	110	135	190

This shows an increase in the volume of air passing the floor brush with 1¼-in. hose, while a higher vacuum is now carried at the hose cock than was necessary when Type A renovator was used in conjunction with the bristle-type of floor renovator. The horse power at the hose cock will now be:

TABLE 13.

HORSE POWER AT HOSE COCK WITH BRUSH TYPE OF BARE FLOOR RENOVATOR OPERATED IN CONJUNCTION WITH TYPE C RENOVATORS.

Size of Hose, In. Diameter.	Length, in Feet.			
	100	75	50	25
	Horse power at hose cock.			
1	1.16	1.32	1.65	2.38
1¼	1.19	1.36	1.60	2.26
1½	1.15	1.32	1.62	2.28

With this combination of floor and carpet renovators, there is no difference in the power consumption when any one of the three sizes of hose is used. However, there is a considerable increase in the quantity of air passing the larger hose. This leads to the statement made by some manufacturers that this increase in air volume results in more efficient cleaning.

Tests given in Chapter III indicate that increase in air volume does not result in any more rapid or efficient cleaning of carpets. The results of actual use of the bare floor brush of the bristle type show no gain when cleaning bare floors. As stated in Chapter IV, the felt-faced renovator, being more effective while it requires less air. In other words, it is the degree of vacuum within the cleaner and not the quantity of air which produces the cleaning in all cases where any degree of vacuum is possible. When intimate contact between the cleaner and the surface cleaned cannot be had, the volume of air determines the efficiency of cleaning. However, the author does not consider that an exhaustion of more than 60 to 70 cu. ft. of free air through cleaners of this type will increase the efficiency to such an extent as to justify the increase of power necessary to adapt a system to larger volumes.

The author considers that with a system in which brushes of the bristle type are to be used, the exhauster should have a capacity of 70 cu. ft. of free air per minute. Such a system is termed by the author a "large volume system," as already mentioned in Chapter IV.

When the felt-covered floor renovator is used instead of the brush, the vacuum within this renovator must not be permitted to rise above 2 in. or the operation of the renovator on the floor will be difficult. To accomplish this, it is necessary to provide openings in the ends of the cleaning slot, as has been explained in Chapter IV. If the vacuum at the hose cock be assumed as 10 in. with 1-in hose, 6 in. with 1¼-in. hose, and 5 in. with 1½-in. hose, and the vacuum within the felt-covered floor renovator be maintained at 2 in. mercury the cubic feet of free air passing the renovator with the various sizes and lengths of hose will be:

TABLE 14.

CUBIC FEET OF FREE AIR PASSING THE FELT-COVERED FLOOR RENOVATORS OPERATED IN CONJUNCTION WITH TYPE A RENOVATORS.

Size of Hose, In. Diameter.	Length, in Feet.			
	100	75	50	25
	Free air, cubic feet per minute.			
1	36	43	54	74
1¼	49	56	68	94
1½	68	78	94	130

These figures show a considerable reduction from those obtained with the brush type of floor renovator, particularly when the larger sizes of hose are used, and considerable reduction can be made in the capacity of the exhauster and still obtain the best results when using carpet renovator and bare floor renovator simultaneously.

The horse power at the hose cock required to operate these felt-faced floor renovators with different sizes and lengths of hose are:

TABLE 15.

HORSE POWER REQUIRED AT HOSE COCK TO OPERATE FELT-COVERED FLOOR RENOVATORS IN CONJUNCTION WITH TYPE A RENOVATORS.

Size of Hose, In. Diameter.	Length, in Feet.			
	100	75	50	25
	Horse power at hose cock.			
1	1.0	1.19	1.49	2.05
1¼	0.72	0.83	1.0	1.39
1½	0.79	0.93	1.13	1.56

In this case, the 1¼-in. hose is the most economical size to use, as was the case with the brush renovators. However, the advantage over the 1½-in. hose is not as great as with the brush renovator.

With this type of renovator, the manufacturer has some control over the length of hose which the operator will use in connection with the bare floor renovator, as he may open the ends of the renovator just sufficiently to produce 2 in. of vacuum under same with, say, 50 ft. of hose. Then, if the operator should attempt to use the renovator with 25 ft. of hose, it will stick and push hard and he will soon learn that a longer hose is necessary.

Conditions for Plant of Small Power.—For locations where it is desirable to sacrifice efficiency somewhat to reduction in the amount of power required, as in residences, the Type A carpet renovator may be used and the vacuum under the same reduced to 2 in. mercury, which will still do effective cleaning, but at a slower rate, as was shown by tests in Chapter III. This requires not exceeding 20 cu. ft. of free air per minute.

With this quantity of air the velocity in the hose must be considered as, in order to have a clean hose at all times, it is necessary to maintain a velocity in the hose of not less than 40 ft. per second. Referring to the diagram, Fig. 48, it will be seen that this velocity will not be obtained in any hose larger than 1¼ in. and this is, therefore, the largest size which can be used. In all the former cases the velocity was so much in excess of this minimum that its consideration was not necessary.

With a vacuum of 2 in. of mercury in the renovator and 20 cu. ft. of air passing, the vacuum at the hose cock will be:

TABLE 16.

VACUUM AT HOSE COCK, WITH 2-IN. VACUUM AT TYPE A RENOVATOR.

Size of Hose. In. Diameter	Length, in Feet.			
	100	75	50	25
	Vacuum at hose cock, in. hg.			
1	4	3.5	3	2.5
1¼	2.6	2.45	2.3	2.15

In this case the increase in vacuum at the renovator would not be objectionable as, with 4 in. vacuum at the hose cock, the vacuum at the renovator would never reach the standard used with the former deductions and the volume of air passing could, therefore, never reach 29 cu. ft. Any increase, due to the use of shorter hose, would, therefore, be an advantage in its approach toward the standard set for the larger plants. Therefore, we will assume that a vacuum of 4 in. mercury will be maintained at the hose cock with 1-in. hose and a vacuum of 2½ in. at the hose cock with 1¼-in. hose.

The renovators for bare floor work will be the felt-covered type and will be opened at the ends just sufficiently to limit the vacuum within the same to 2 in. mercury when operating with 25 ft. of hose. This will require the passage of 40 cu. ft. of free air per minute when 1-in. hose is used and 35 cu. ft. when 1¼-in. hose is used. The horse power at the hose cock will be 0.39 H. P. with the 1-in. diameter hose and 0.17 H. P. with the 1¼-in. hose. Here again we see that the 1¼-in. hose is the more economical to use.

If bristle brushes are used with this system at the same time that carpet renovators are in use, the quantity of air which will have to pass them, in order to maintain the vacuum on the system at the proper point to do effective cleaning with the carpet renovators, will be:

TABLE 17.

AIR QUANTITIES WHEN BRISTLE BARE FLOOR RENOVATORS ARE USED IN CONJUNCTION WITH TYPE A CARPET RENOVATORS AT 2 IN. HG.

Size of Hose, In. Diameter.	Length of Hose, in Feet.			
	100	75	50	25
	Free air, cu. ft. per minute.			
1	30	36	42	60
1¼	41	48	60	80

The use of these brushes in plants of more than one-sweeper capacity would require the use of an exhauster of greater capacity than is required for either the carpet or the bare floor renovator. Where the plant is of but one-sweeper capacity, the quantity of air that would pass these brushes, were the plant of proper capacity to serve the carpet and floor renovators, would not be sufficient to do effective work, as was explained in Chapter IV. In such cases, this arrangement should be prohibited.

A system of the type just described is what has been termed by the author as a "small volume" plant in Chapter IV.

Limit of Length for Hose.—The author has made the deductions in this chapter, using 100 ft. of hose as the maximum length. This is considered to be the greatest length that should be used. The adoption of a shorter length is recommended by many manufacturers, but the author does not consider that the advantage to be obtained by the adoption of a shorter length justifies the additional expense of piping which will result in many cases. This will be governed by the character of the building and, in many cases, it will be possible to use 50 ft. as a maximum. It has been the practice of the author to lay out his installations so that any point on the floor of any room may be reached in the most direct line with 75 ft. of hose. When this is done 100 ft. of hose will easily clean any part of the walls or ceilings and give an ample allowance for running around furniture or other obstructions.

The figures in this chapter will demonstrate to the reader the part that the cleaning hose plays as a limiting factor in the operation of a vacuum cleaning system and shows the care that must be exercised in the selection of the proper hose for each condition.

CHAPTER VII.

Pipe and Fittings.

As we continue to follow the dust-laden air in its passage toward the vacuum producer we next encounter that portion of the conduit which is permanently and rigidly fixed in place in the building; namely, the pipe line, its fittings and other appliances.

Hose Inlets.—The first portion of this conduit which we must consider is the point where the hose is attached to the pipe line; that is, the inlet, or, as it is often improperly termed, the "outlet" valves.

As it is necessary to close the inlets air tight when they are not in actual use, in order to prevent the entrance of air except through the hose lines in use, some kind of a cut-off valve must be provided, as well as a receptacle into which the end of the hose may be connected when desired.

With the earlier systems a high degree of vacuum was carried in the pipe lines and the vacuum producers were of small displacement. Slight leakage would greatly reduce the capacity of the system and the best form of valve was necessary. The valve adopted was the ordinary ground-seat plug cock, on account of its unobstructed air passage and air-tight closing. The hose was connected to these cocks either by a ground-joint, screwed coupling or by a slip coupling similar to those used to unite the sections of the cleaning hose. An inlet cock of this type is illustrated in Fig. 51.

These cocks projected about 4½ in. beyond the face of the finished wall and formed a considerable obstruction, especially when located in halls or corridors. In order to reduce the projection into the apartment the manufacturers of the systems using screwed-hose couplings and substituted a projecting nipple closed by a cap screwed in place. The whole projected only ¾ in. beyond the finished wall line.

These outlets were suitable for use only with hose having screwed connections. When an attempt is made to remove the cap with the vacuum producer in operation, there is a tendency for the vacuum to cause the cap to hug the last thread and render its removal difficult. Also, when the suction is finally broken it is accomplished with considerable hissing noise.

In order to permit the use of the slip type of hose coupling, a hinged flap valve was substituted for the screwed cap, a rubber gasket being placed under the cap. This was held firmly in place by the vacuum in the pipe line. The interior of the casting inside of the flap was turned to a slip fit for the end of the hose coupling. With this type of valve and the slip hose coupling, described in Chapter VI, it is possible to reverse the hose to equalize wear and remove obstructions.

These inlets have been made with valves that are closed only by gravity when there is no vacuum on the system and many are so constructed that when opened wide they will remain open with the vacuum on the piping. This type of valve

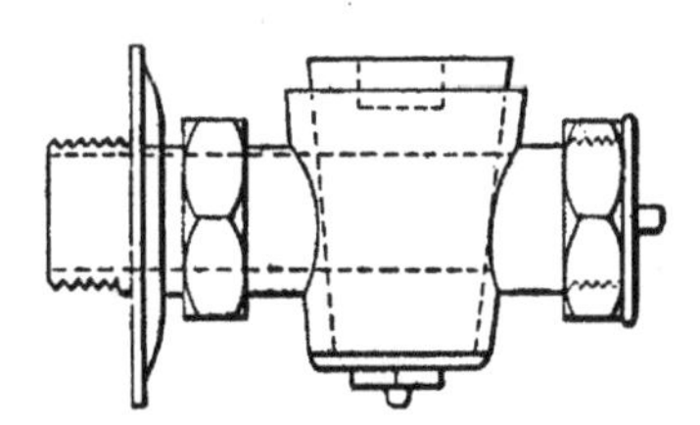

FIG. 51. INLET COCK TO PREVENT AIR LEAKAGE WHEN NOT IN USE.

will often be opened by the inquisitive person when no vacuum exists in the system and as there are no immediate results, they may be left open with the result that there will be a very large leakage of air on starting the vacuum producer. This makes it necessary for some one to make a tour of the building in order to close the valve which is open before the system can be efficiently operated. If the vacuum producer is designed to operate several renovators simultaneously, it may not be discovered that there are any valves open and a considerable amount of power will be wasted.

In order to overcome this difficulty it is necessary to provide a spring on the hinge of the flap valve that will automatically close the valve whenever the hose is withdrawn. When the inlets are located in public places they should be fitted with a lock attachment to prevent them from being opened by unauthorized persons.

A valve of this type is illustrated in Fig. 52. This valve has a projection on its inner face which engages with a ridge on the hose couplings, preventing the removal of the hose without slightly raising the cap and making it impossible to accidently pull the hose out of the inlet.

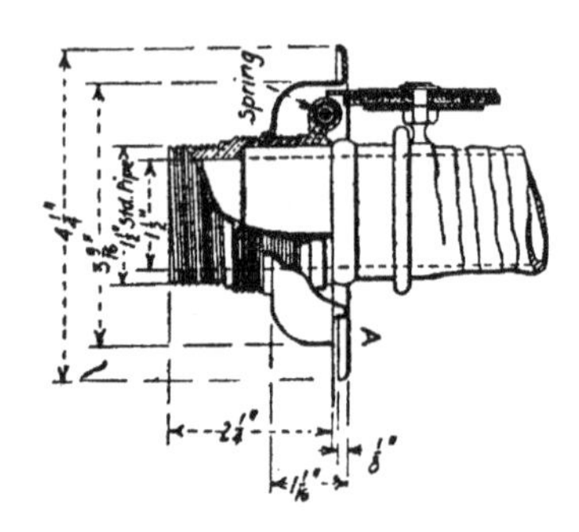

FIG. 52. TYPE OF AUTOMATIC SELF-CLOSING INLET COCK.

The particular valve here shown is suitable for use only with the all-rubber hose connection described in Chapter VI.

We must next consider the material of which the conduit itself is to be made. The commercial wrought-iron or mild steel, screw-jointed pipe, such as is used for water and steam lines, is probably the best suited for this purpose and was the first material used. In earlier installations the pipe was galvanized, but, owing to the tendency for the zinc coating to form irregularities within the pipe, its use has been abandoned in favor of the commercial black iron pipe.

Seamless drawn tubing would undoubtedly make the ideal material for this purpose. However, the ordinary butt or lap-welded pipe is satisfactory and is now generally used.

Sheet metal pipe was introduced by one manufacturer but its use was shortly abandoned in favor of the commercial pipe.

As joints and changes in direction are necessary in the pipe lines, some sort of fittings must be used. The ideal conduit for passage of dust-laden air should be of uniform bore and as smooth on the inside as a gun barrel. Various attempts have been made to accomplish this result in commercial installations, one of which is illustrated in Fig. 53. These fittings are made up of three parts for a coupling and four for a branch or change in direction. One of these is screwed on to the end of each piece of pipe, the pipe butting against a shoulder and the end of the pipe made to register with the bore of the fitting by reaming. This piece is faced true and fitted against the face of the casting, forming the bend or branch, or fitted against the piece on the end of the other length of pipe. A thin gasket is placed between them, a projecting ring on one piece fitting

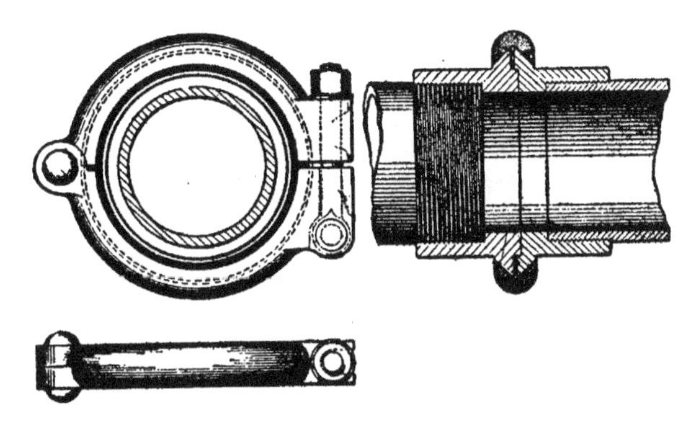

FIG. 53. "SMOOTH BORE" PIPE COUPLING.

into a groove on the other, causing the bore of the two halves to register. The two halves are joined together by the V-grooved clamp, held in place by a small bolt. This is theoretically an ideal joint, but the clamp is not of sufficient strength to withstand the strain of settlement of the building and breakages are frequent. Several instances of this character, particularly on steamers, have come to the observation of the author, and there are several buildings which have been roughed in with this type of fitting, used on concealed piping, which were found to be useless on the completion of the building, due to breaking of the joints in inaccessible places.

A modification of this joint which will have ample strength can be made by using standard pipe flanges, screwing the pipe through the flange and facing the end off in a lathe. Fittings could be made with a bore equal to that of the pipe and proper alignment secured by the use of dowel pins, as illustrated in Fig. 54. The cost of making this joint would be high and they would occupy too much space to be easily concealed in partitions, furring or other channels usually provided for the reception of such piping.

The standard Durham recessed drainage fitting, having the inside cored to the bore of the pipe and recesses provided for the threads as used in connection with the modern plumbing system, if left ungalvanized and having the inside well sandblasted to remove all rough places, makes a serviceable fitting. Care should be exercised to cut the threads on the piping of

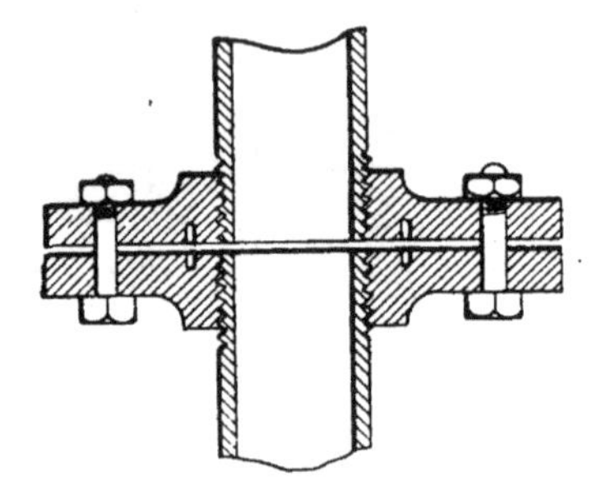

FIG. 54. JOINT MADE OF STANDARD PIPE FLANGES.

proper depth to allow the end of the pipe to come as close to the shoulder of the recess as practicable and to obtain a tight joint. The end of the pipe should be carefully reamed before assembling.

These fittings have become standard with nearly all manufacturers and are illustrated in Fig. 55, which shows the right and wrong way to install same.

Trouble was experienced on some of the earlier systems using high vacuum with the fittings cutting out on the side subjected to the impact of the dust-laden air. To overcome this trouble one manufacturer re-inforced the fittings by increasing the thickness of metal at the point affected. The trouble was undoubtedly caused by too high a velocity in the pipe line, as in the case of the small brass stems, explained in Chapter V. With the introduction of vacuum control and larger pipes, this trouble disappeared and the special fittings never came into general use.

While the utmost care should be taken to prevent stoppage of the pipe lines these stoppages are likely to occur in the best-

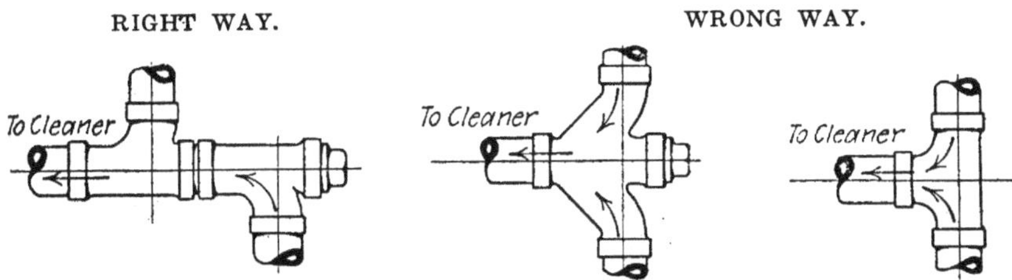

Use two Y-branches instead of straight or cleanout tees. In case the latter are used the dirt will shoot by into the other branch.

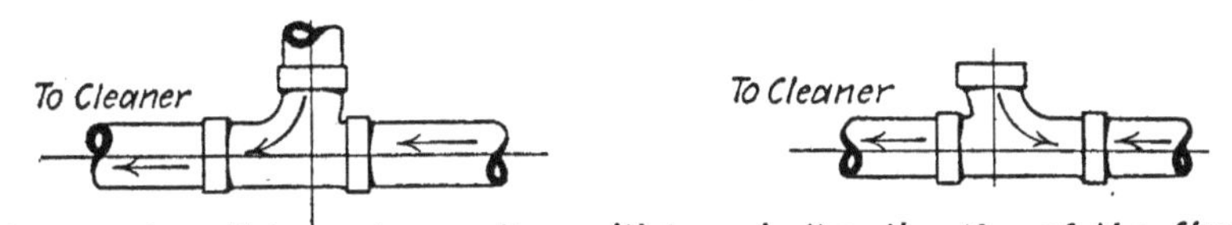

Always place Y-branches so they will turn in the direction of the flow.

Place the clean-out at right angles to the direction of flow entering the fitting. Otherwise it serves as a pocket to catch passing dirt.

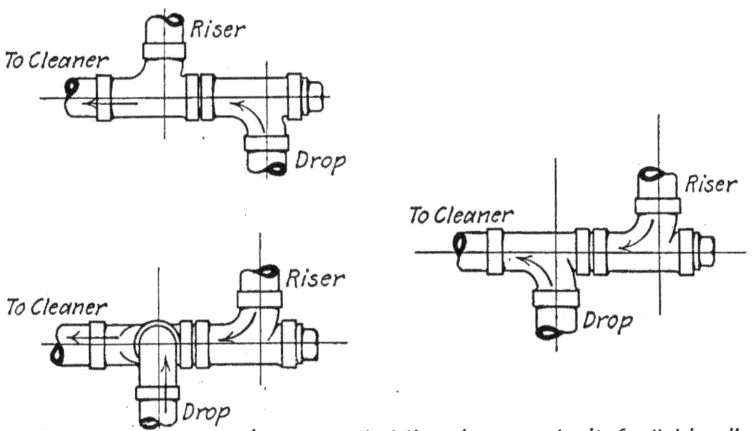

Special care must be exercised to see that there is no opportunity for dirt to collect in the basement drops. Above is shown a common wrong way and two possible right ways.

FIG. 55. STANDARD DURHAM RECESSED DRAINAGE FITTINGS GENERALLY USED IN VACUUM CLEANING INSTALLATIONS

constructed lines and ample clean-out plugs should be provided for the removal of such stoppage. Brass plugs are the most serviceable for this purpose, as they are easily removed when necessary and can usually be replaced air tight.

The brass clean-outs, while most satisfactory, are costly when installed in large sizes. Equally satisfactory results can be obtained at a lower cost by using 2-in diameter plugs on all lines 2 in. and over in diameter.

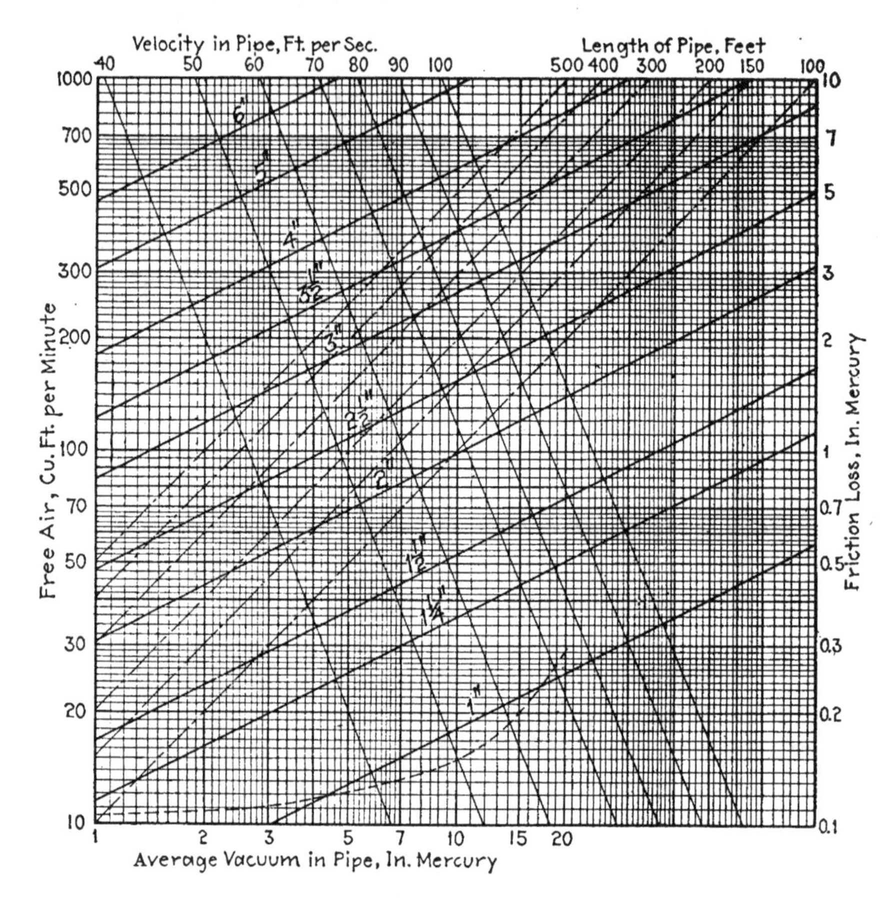

FIG. 56. FRICTION LOSS IN PIPE LINES.

Matches are perhaps the most frequent cause of stoppage in pipe lines. Stoppage from this cause can be largely avoided by the use of pipe of sufficient size to permit the match to turn a complete somersault within the pipe whenever it catches against a slight obstruction or rough place in the pipe or fittings. A 2-in. diameter pipe is just large enough to permit this and smaller sizes of pipe should be avoided whenever possible.

Pipe Friction.—The friction loss in piping follows the same law as that in hose lines and is easily computed by use of the chart (Fig. 56), which is constructed on the same general principle as the chart of hose friction (Fig. 48). The directions for use of the hose chart apply to the pipe chart. In computing this chart the actual inside diameter of the commercial wrought-iron pipes have been used instead of the nominal diameters, resulting in an increased capacity for all sizes except 2½-in. which is less than the nominal diameter.

Determination of Proper Size Pipe.—Friction in the pipe lines tends to increase the vacuum to be maintained and therefore the power to be expended at the vacuum producer and should be kept as low as possible. The pipe sizes should be made as large as conditions will permit. The limit of size is fixed by the velocity in the pipe. When it is necessary to lift the dirt to any extent, the velocity should not be allowed to fall below 40 ft. per second at any time. When the pipe is a vertical drop, the velocity does not matter as gravitation will assist the air current in removing the dirt. When the line is horizontal a lower velocity than 40 ft. per second is permissible at times, provided that this minimum velocity is exceeded at frequent intervals to flush out any dirt that has lodged in the pipe during periods of low velocity.

If a Type A renovator is used with 1-in. hose and a vacuum of 10 in. of mercury maintained at the hose cock, the minimum air passing, with 100 ft. of hose in use, will be 29 cu. ft. of free air per minute, which is equivalent to 44 cu. ft. at 10 in. of vacuum. The entering velocity in the pipe should be calculated with air at this density. This will give a velocity of 50 ft. per second in a 1½-in. pipe, but only 30 ft. per second in a 2-in. pipe. Therefore, the 1½-in. pipe is the largest that should be used where lifts occur on a line serving but one Type A renovator with 1-in hose. When the renovator is tilted at a considerable angle or lifted from the carpet, as will frequently occur in cleaning operations, the quantity of air passing the renovator will be upwards of 42 cu. ft. of free air, equivalent to 62 cu. ft. at 10-in. vacuum. When this occurs the velocity in a 2-in. pipe will be 44 ft. per second, which will be ample to flush a horizontal line of piping.

If 1¼-in hose is used with a Type A renovator, the minimum quantity of air will be 29 cu. ft. and the vacuum entering the pipe will be 6 in. mercury, giving an equivalent volume of 37 cu. ft. This will produce a velocity of 42 ft. per second in a 1½-in. pipe, which is the largest that can be used where a lift occurs. However, when the renovator is lifted free of the carpet, the air quantity will be 62 cu. ft. of free air, equivalent to 80 cu. ft. at 6 in. of vacuum, and will produce a velocity of 39 ft. per second in a 2½-in. pipe. This would be just about sufficient to flush a horizontal line.

If 1½-in. hose were used the air quantity will be 29 cu. ft. and the vacuum entering the pipe 5 in. mercury, equivalent to 35 cu. ft. This will give a velocity in a 1½-in. pipe of 40 ft. per second. When the renovator is raised from the carpet, the air quantity will be upwards of 90 cu. ft. of free air, equivalent to 110 cu. ft. at the density of that entering the pipe, and will produce a velocity of 33 ft. per second in a 3-in. pipe. This is too low to thoroughly flush a horizontal pipe.

The figures given above are repeated from Chapter VI and show that the use of 1¼-in. instead of 1-in. hose, permits the use of a larger-sized horizontal pipe line for serving one renovator, but that the use of 1½-in. hose, instead of 1¼-in., will not permit of any enlargement in the pipe size. Since we have seen in Chapter VI that a 1¼-in. hose gives the least expenditure of power when used with a Type A renovator, there will be no gain from a reduction in the pipe friction due to the adoption of this hose.

The dependence on the raising of the renovator from the floor to flush out a larger pipe line should not be carried beyond that to be obtained from a single renovator. That is, when the pipe must serve more than one renovator at the same time, the quantity of air that two or more renovators will pass, if they were raised from the floor at the same time, should not be used in determining the limiting velocity in the pipe, as such an occurrence is not likely to be obtained often enough to thoroughly flush the pipe. Furthermore, there will be times when this pipe will have to serve only one renovator and the pipe will not be adequately flushed. When the pipe is serving more than one renovator, the actual air passing the renovators should be used in determining the maximum size of pipe and it is advisable to use this maximum size in nearly all cases where the structural conditions will permit.

These sizes will then be:

TABLE 18.

PIPE SIZES REQUIRED, AS DETERMINED BY AIR PASSING RENOVATORS.

Number of Renovators in Use.	Cu. Ft. per min.	Pipe Sizes, In. Diam.	
		With 1-in. Hose.	With 1¼-in. Hose.
1	29	2	2½
2	58	2½	2½
3	87	3	3
4	116	3½	3½
5	145	3½	3½
6	174	4	4

Using these maximum sizes, the friction loss in a pipe line, with carpet renovators in use exclusively, will be:

TABLE 19.

FRICTION LOSS IN PIPE LINES, WITH CARPET RENOVATORS IN USE EXCLUSIVELY.

Number of Sweepers.	Friction Loss per 100 Feet, Inches.	
	With 1-in. hose.	With 1¼-in. hose.
1	0.20	0.06
2	0.30	0.20
3	0.24	0.17
4	0.19	0.13
5	0.30	0.22
6	0.24	0.17

These friction losses are figured with a density of air in the pipe equal to 6-in. vacuum in case of the 1¼-in. hose and 10-in. vacuum in case of the 1-in. hose, which will be the density of the air entering the pipe, while the average density should be used in order to give correct results. If the pipe line is not over 400 ft. equivalent length the results will be approximately correct.

These results show, first, that the friction loss in pipe lines is much lower than that in the hose lines used with the same system; second, that the higher vacuum in the pipe causes greater loss, an argument in favor of the use of larger hose.

These friction losses are obtained only when carpet renovators are used exclusively and all the renovators are held in the proper position to perform the most economical cleaning. In actual practice this condition will not exist except when one renovator is used. Where more than one renovator is in use simultaneously, some of the renovators will be raised from the floors at the time others are in position to do effective cleaning

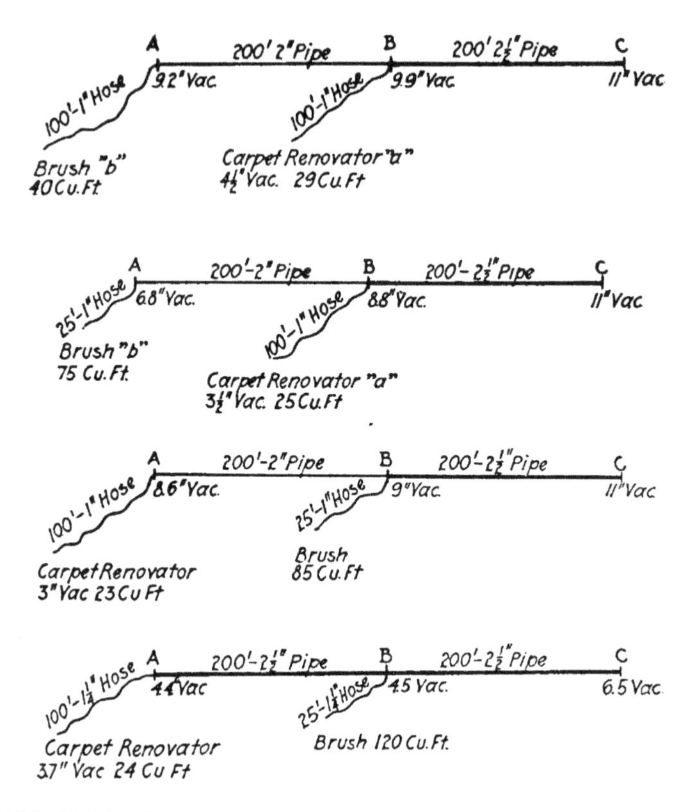

FIGS. 57-60. DIAGRAMS SHOWING OPERATION OF BRUSH AND CARPET RENOVATORS UNDER DIFFERENT CONDITIONS.

and will admit a greater quantity of air, increasing the friction. This is not a serious condition as the time that the renovators will be raised is only a small part of the total time spent in cleaning and will merely reduce the efficiency of the other renovators temporarily. However, when brushes or floor renovators are used at the same time as the carpet renovators, there will be a continuous flow of air in greater quantities through these brushes, which will permanently increase the friction loss. The use of a single brush or floor renovator with the same sized pipe as is necessary to operate the carpet renovator will not reduce the efficiency of the brush, as a high degree of vacuum at the brush or floor renovator is not necessary or even permissible and a further slight reduction will not affect the operation of these renovators.

When a brush or floor renovator is used on the furthest outlet from the vacuum producer at the same time that carpet renovators are being used on outlets nearer the vacuum producer, the larger quantity of air passing the brush will tend to reduce the vacuum at the hose cock to which the carpet renovator is attached and thereby impair its efficiency. For example, if we have a brush renovator connected through 100 ft. of 1-in. hose to an outlet at the end of a pipe line 400 ft. long, properly designed to serve two carpet renovators, the vacuum at the separators should be maintained at 10-in., plus 2×0.20 plus 2×0.30, or 11 in. of mercury. Suppose that this vacuum is automatically maintained at this point and a carpet renovator be attached 200 ft. from end of pipe (Fig. 57). The quantity of air passing through the 2½-in. pipe B-C will be approximately 29 plus 40 or 69 cu. ft., and the friction loss in this pipe will be 1.1 in. The vacuum maintained at the outlet B (Fig. 57) will be 9.9 in. or approximately the correct vacuum to maintain 4½-in. vacuum at the renovator "a." The friction loss in the pipe line from B to A will be 0.7 in. and the resulting vacuum at the hose cock A will be 9.2 in. The quantity of air passing the brush will be 40 cu. ft. Under these conditions there will be no loss in efficiency of cleaning due to the brush renovator being used on the end of the line. If the operator using the brush at the outlet A should use only 25 ft. of hose instead of 100 ft. (Fig. 58) the air passing this brush will be 75 cu. ft. and the vacuum at the hose cock A will be 6.8 in. The vacuum at the hose cock B will be 8.8 in. and the vacuum at the carpet renovator "a" will be reduced to 3½ in. with 25 cu. ft. of air passing, which will reduce the efficiency of the carpet renovator "a."

If the brush renovator be attached to the hose cock B (Fig. 59), using 25 ft. of hose, the vacuum at hose cock B will be

9 in. and the brush renovator will pass 85 cu. ft. of air, while the vacuum at hose cock A will now be reduced to 8.6 in. and the vacuum at the renovator will be reduced to 3 in. mercury and the air passing to 23 cu. ft.

If a brush type of renovator be used at each outlet. with 25 ft. of hose in each case and the vacuum at the separator be maintained at 11 in. mercury the vacuum at hose cock B will be 7 in. and brush "a" will pass 76 cu. ft. of air while the vacuum at hose cock "a" will be 5 in. and brush "b" will pass 63 cu. ft. of air or a total of 144 cu. ft., which will be in excess of the 70 cu. ft. per renovator. recommended as the capacity of the plant in Chapter VI. This will not result in any loss of efficiency if the vacuum producer be designed to handle but 140 cu. ft. as a maximum, for the vacuum at the separator will then fall to a point where but 140 cu. ft. passes, resulting in a decrease in the vacuum throughout the system. But as only brushes are now in use there will be no loss in efficiency, owing to the reduction in the vacuum at the brushes.

When 1¼-in. hose is used with a carpet renovator at the end of the pipe line connected through 100 ft. of hose and a brush at the hose cock B connected through 25 ft. of hose (Fig. 60), the worst case of the three already cited, the vacuum at the separator being maintained at that necessary to carry 4½ in. when two carpet renovators are in use, the vacuum at the hose cock B will be 4.5 in. and brush "a" will pass 116 cu. ft. of air while the vacuum at hose cock A will be 4.4 in. and the vacuum in renovator "b" will be 3.7 in and will pass 24 cu. ft. of air.

These are better cleaning conditions than were obtained when 1-in hose was used. It will be noted that the total air passing the exhauster is now 140 cu. ft. and this must not be reduced or there will be a falling off in the vacuum at the carpet renovator "b." It is, therefore, necessary for the exhauster to be capable of handling 140 cu. ft. of air or 70 cu. ft. of air per renovator in order to do effective carpet cleaning when carpet renovators and brushes are used in conjunction.

When two floor brushes are used with the above arrangement of pipe and hose, the vacuum must fall considerably or the air quantity be greatly increased. However, the reduction in vacuum will not result in serious loss in efficiency when only brushes are in use.

When a larger number of sweepers are used with a system of piping, it is necessary to allow 70 cu. ft. of free air per sweeper in figuring the sizes of pipe to be used, and the total loss of pressure in the piping between the outlet farthest from the vacuum producer and that nearest to same must be limited in order to prevent too wide a difference in the vacuum at the hose cock when all the sweepers for which the plant is designed are in use. The author considers that this loss in pressure should not be greater than 2 in. mercury in order to give satisfactory results.

Before the piping system can be laid out and the sizes of piping determined it is necessary to ascertain, first, the number of sweepers to be operated simultaneously and the number of risers necessary to properly serve these sweepers.

Number of Sweepers to be Operated.—This is determined by the character of the surfaces to be cleaned, the amount of such surface, and the time allowed for cleaning.

It has been demonstrated in actual practice that one operator can clean as high as 2,500 sq. ft. of carpet when same is on floors of comparatively large areas, and not over 1,500 sq. ft. when the carpets are on small rooms; 2,000 sq. ft. is considered to be a fair average.

Bare floors are cleaned more rapidly. In school house work an ordinary class room has been cleaned in 10 minutes, or at the rate of 7,200 sq. ft. per hour, but time is occupied in moving from one room to another and the writer considers 5,000 sq. ft. per hour as rapid cleaning and 3,500 sq. ft. as a fair average.

The time of cleaning will vary in buildings of different character and used for different purposes. In office buildings the cleaning force work throughout the night or about 10 hours, while in school houses the cleaning is done by the janitor force which has been on duty throughout the school period and the time is necessarily limited to about two or three hours after school hours, the corridors and play rooms being cleaned during the school period and only the class rooms being cleaned after closing time.

Let us assume, as an example, an office building having eight floors each 100 ft. x 150 ft., with a floor plan as shown in Fig. 61.

The corridors, stairs and elevator halls will probably be floored with marble which must be scrubbed in order to remove the stain accumulated during the day and they will not be considered in connection with a dry vacuum cleaning system. The area of the floors in the offices on any floor will be approximately 10,000 sq. ft. and one floor can be cleaned by one operator in 5 hours, or two floors during the cleaning period, so the plant must be of sufficient size to serve four sweepers simultaneously.

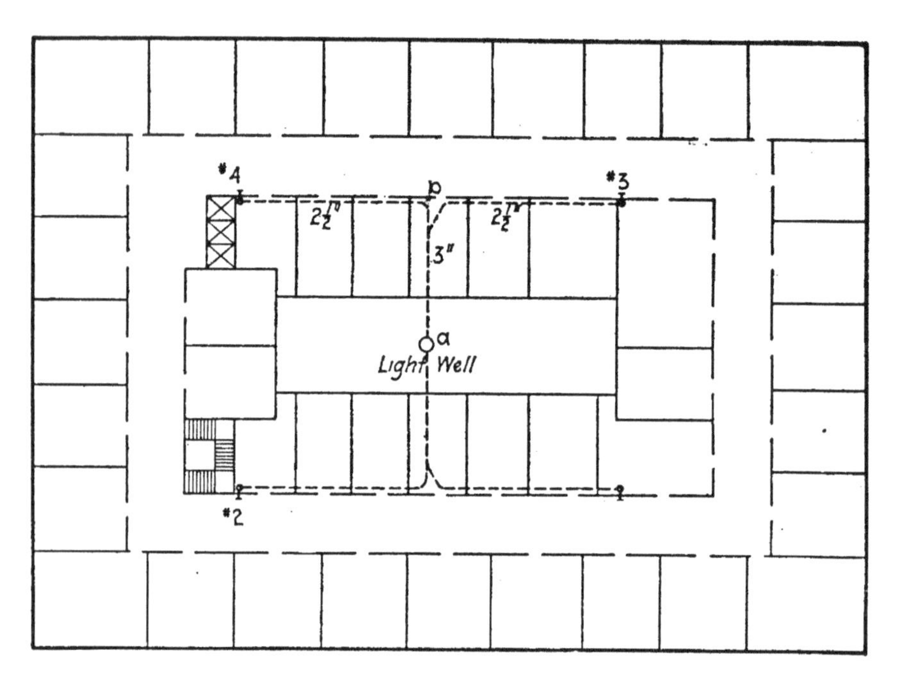

FIG. 61. TYPICAL FLOOR PLAN OF OFFICE BUILDING ILLUSTRATING NUMBER OF SWEEPERS REQUIRED.

In a school house containing four class rooms, where the janitor cleans the play rooms and corridors during the school period, as can be readily done with a vacuum cleaner since there will be no dust scattered about to fill the air and render it unsanitary, the class rooms can easily be cleaned in one hour by one operator. The author considers that one sweeper capacity for each six to eight rooms is ample for a large school.

Buildings of special construction and used for special purposes must be considered differently according to the conditions to be met, but the size of the plant can be readily determined in each case by use of the rules already given.

Number of Risers to be Installed.—Much difference of opinion exists among the various manufacturers of vacuum cleaning systems as to the maximum length of hose that should be used with a cleaning system, and as this maximum length determines the number of risers to be installed, some fixed standard is necessary. As already stated in Chapter VI, the author considers that this maximum should be fixed at 75 ft.; that is, the risers should be so spaced that all parts of the floor of the building can be reached with 75 ft. of hose. Where 50 ft. is used as a maximum, as is recommended by many manufacturers, the number of risers would be increased, incurring a greater cost of installation and requiring the operator to shift his hose from one inlet to another more often than would be the case where fewer inlets were used, and more time would be required in cleaning, with a slight reduction in the power. The author does not consider that this reduction in power would be sufficient to offset the additional time required to change the hose from one inlet to another.

The best and quickest way to determine the number of risers necessary is to cut a piece of string to the length representing 75 ft. on the scale of the plans, and by running this around the plan using corridor doors for access to all rooms, wherever possible, locate the riser so that every point can be reached with the string. In the case of the building illustrated in Fig. 61 four risers located as shown will be necessary.

Size of Risers.—Before we can determine the size of risers to be installed it is necessary to determine the probable number of sweepers that will be attached to any one riser simultaneously. In the case of the building (Fig. 61) it is possible that there may be four sweepers attached to one riser and it is also possible that there may be but one, and two sweepers to a riser is considered to be a safe assumption. The author uses the following rule in determining the size of risers to use:

Where the number of sweepers is double the number of risers, assume that all sweepers will be on one riser simultaneously.

Where the number of sweepers is equal to the number of risers, assume that half the sweepers will be on one riser simultaneously.

Where the number of sweepers is half the number of risers, assume that one-quarter of the sweepers will be on one riser simultaneously.

When no lifts occur a low velocity in the riser is not objectionable and the size of the riser should be made equal to the size of the horizontal branch thereto throughout its length, wherever this branch is not larger than 2½ in. diameter. When larger, reductions in the riser can be made until 2½ in. is reached when this size should be maintained throughout the remainder of its length. No riser should be made less than 2½ in. unless a lift is necessary.

Before finally fixing the size of riser to be used in any case the size of the branch in the horizontal lines serving the same must be approximately determined.

These sizes will be dependent on the location in which it is necessary to install the vacuum producer. In the case of the building (Fig. 61) the most desirable location for the vacuum producer will be in the exact center of the building.

With the vacuum producer centrally located the longest run from any riser will be 55 ft. To this we must add:

5 ft. for each long-turn elbow.

10 ft. for each short-turn elbow.

10 ft. for entrance to each long sweep Y branch.

20 ft. for entrance to a tee branch, except at sweeper inlets on risers, where 10 ft. is ample

In calculating the riser friction for risers under 150 ft. in length the whole capacity of the riser can be assumed as being connected to a point midway of its length.

In the eight-story building (Fig. 61) the length of the riser from basement ceiling to eighth floor will be 100 ft. and the length to be figured, 50 ft. The equivalent length of pipe line for any of the risers, with the vacuum producer centrally located, will be:

From entrance tee into riser	10 ft.
Length of riser, one-half total length	50 ft.
Turn at base of riser	10 ft.
Run in basement	55 ft.
Y branch or elbow	10 ft.
Elbow at separator	5 ft.
Equivalent length	140 ft.

Each riser is to serve two sweepers and must pass 140 cu. ft. of free air per minute. This will give a friction loss in a 2½-in. pipe of 2 in. mercury, if 10 in. mercury be maintained at the hose cock and 1-in. hose used; and 1.5 in. mercury if 6 in. mercury be maintained at the hose cock and 1¼-in. hose used. Either of these figures are within the limits set for the maximum friction loss and 2½-in. pipe will be the proper size for the risers and their branches in the basement.

The portion of the main in the basement that serves the two risers on either side of the building (portion "ab," Fig. 61) must be of such size as will produce the same loss in vacuum with 280 cu. ft. of air passing as the 2½-in. pipe gives with 140 cu. ft. of air passing. This may be determined from any table of equalization of pipes or may be obtained from the chart, Fig. 48, in the following manner:

Find the intersection of the horizontal line "140" with the diagonal representing a 2½-in. pipe and pass on the nearest vertical to its intersection with the horizontal line "280." The diagonal inclined toward the left passing nearest this intersection will be the pipe size required. In this case a 3-in. pipe will give a slightly greater friction and will be sufficient.

Unfortunately, it is rarely possible to locate the vacuum producer in as favorable a point as that given in the illustration, but an effort should always be made to select a location as nearly central to all risers as possible. The basements of modern office buildings are generally crowded and the space assigned to the mechanical equipment is limited and owing to the necessity of ventilation, the vacuum producer is generally located near the outside of the building.

Probably the best location that could be obtained in this case would be at "d" (Fig. 62). The length of piping to risers 1 and 2 would now be the same as that to all risers in case of Fig. 61, but the distance to risers 3 and 4 will be increased 50 ft. It will be possible to increase the size of the pipe line "bd" to the maximum size to serve four sweepers, or 3½ in., the risers and their branches to remain 2½ in.

The total friction loss to risers 1 and 2 will now be:

Entrance to tee in risers, 10 ft. plus 50 ft.	60 ft.
Turn at base of riser, 10 ft., branch from "c" to riser 32 ft.	42 ft.
Entrance to tee in main	20 ft.
Total equivalent length of 2½-in. pipe	122 ft.

When 1-in. hose is used the density of the air entering the 2½-in. pipe is equivalent to a vacuum of 10 in. mercury and the friction loss in the 2½-in. pipe will be 1.9 in. mercury.

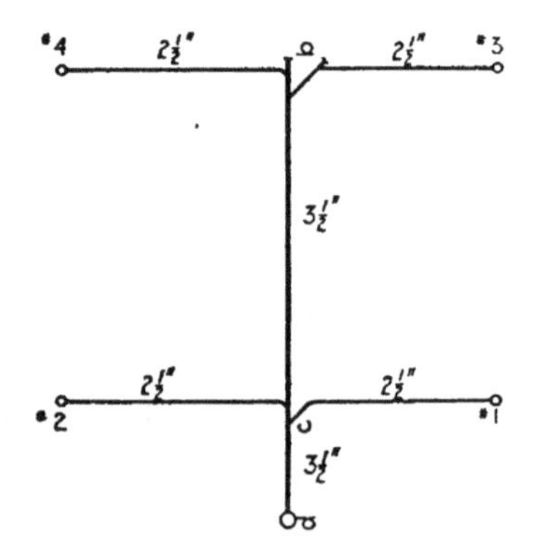

FIG. 62. ELEVATION OF LAYOUT FOR OFFICE BUILDING, SHOWING BEST LOCATION (AT D) FOR VACUUM PRODUCER.

When 1¼-in. hose is used, the density of the air entering the pipe will be equivalent to a vacuum of 6-in. mercury and the friction loss in the 2½-in. pipe will be 1.32 in. mercury.

The density of the air entering the 3½-in. pipe, "b d," will be equivalent to a vacuum of 11.9 in. mercury when 1-in. hose is used, and to 7.32 in. mercury when 1¼-in. hose is used. The friction loss in the 3½-in. pipe will be 0.31 and 0.23 in.

mercury, respectively. Total friction loss to inlets on risers 1 and 2 will be 2.21 in. with 1-in. hose in use, and 1.55 in. with 1¼-in. hose.

To obtain the friction loss to inlets on risers 3 and 4 the friction loss in the pipe "bc" must be added to the above figures. With 50 ft. of 3½-in. pipe carrying 280 cu. ft. free air the friction loss is 0.6 in. when the vacuum in the pipe is 12 in. and 0.4 when the vacuum in the pipe is 8 in.

The total loss of vacuum to inlets on risers 3 and 4 will be 2.91 in. if 1-in. hose is used and 1.95 in. if 1¼-in. hose is used. In this case, the total loss from inlet to vacuum producer is approximately equal to the maximum variation of vacuum permitted at sweeper outlets when 1¼-in. hose is used, but is greater than when 1-in. hose is used.

However, it is the variation in vacuum at the hose cock farthest from and that nearest to the vacuum producer that fixes the maximum variation allowable. In this case it will be the difference in vacuum between an inlet on riser 1 or 2 and a similar inlet on riser 3 or 4. The difference in vacuum at the bases of these risers will be the friction loss in the pipe "bc," and the total difference in friction in the risers will occur when one sweeper is attached to the lowest inlet on one riser, and one sweeper on the eighth and one on the seventh floor on the other riser. The friction loss in the riser having the two sweepers attached to its upper inlets will be:

15 ft. of 2½-in. pipe from seventh to eighth floors, 70 cu. ft. of free air per minute, or 0.051 in. with a density equivalent to 6-in. vacuum, and 0.075 in. with a density equivalent to 10-in. vacuum.

85 ft. of 2½-in. pipe from first to seventh floors, 140 cu. ft. free air per minute, or 0.25 in. with a density equivalent to 6-in. vacuum, and 0.42 in. with a density equivalent to 10-in. vacuum.

The total difference in vacuum at the hose cocks will be:

$0.051+0.25+0.4=0.7$ in. with 6-in. vacuum at the hose cock.

$0.075+0.42+0.6=1.15$ in. with 10-in. vacuum at the hose cock.

Either of these values are well within the maximum variation. It is, therefore, evident that when the vacuum producer cannot be centrally located that a piping system which will give the most nearly equal length of pipe to each riser will yield the best results.

A vacuum cleaning system for serving a passenger car storage yard will best illustrate the effect of long lines of piping. A typical yard having 8 tracks, each of sufficient length to accommodate 10 cars, is shown in Fig. 63. The vacuum producer in this case is located at the side of the yard at one end, which is not an unusual condition.

The capacity of this yard will be 80 cars which must generally be cleaned between the hours of midnight and 6 A. M., or a period of 6 hours for cleaning.

It will require one operator approximately 20 minutes to thoroughly clean the floor of one car, on account of the difficulty in getting under and around the seat legs. In addition to this, it is also necessary to clean the upholstery of the seats and their backs, which will require approximately 25 minutes more or 45 minutes for one operator to thoroughly clean one car. Therefore, one operator can clean 8 cars during the cleaning period and a ten-sweeper plant will be necessary to serve the yard.

One lateral cleaning pipe must be run between every pair of tracks or four laterals in all to properly reach all cars without running the hose across tracks where it might be cut in two by the shifting of trains.

Outlets should be spaced two car lengths apart in order to bring an outlet opposite the end of every second car. This will make it possible to bring the hose in through the end of the car at the door opening and clean the entire car from one end which can be done by using 100 ft. of hose. The use of double the number of outlets and 50 ft. of hose would require two attachments of the hose to clean one car resulting in a loss of time in cleaning and is not recommended.

In this case, 100 ft. of hose would be the shortest length that would be likely to be used and 60 cu. ft. of free air would be the maximum to be allowed for when using 1¼-in. hose.

The simplest layout for a piping system to serve this yard would be that shown in Fig. 63.

When the entire yard is filled with cars and the entire force of ten operators is started to clean them it would be possible to so divide them that not over three operators would be working on any one lateral and this condition will be assumed to exist. The maximum size for the laterals between the tracks will be that for three sweepers, or 3 in., and it will not be safe to use this size beyond the second inlet from the manifold, from which point to the end of the lateral it must be made 2½ in., the maximum size for either one or two sweepers. The total loss of pressure due to friction from the inlet at x (Fig. 63) to the separator can be readily calculated from the chart (Fig. 56) as follows:

TABLE 20.

PRESSURE LOSSES FROM INLET TO SEPARATOR IN SYSTEM FOR CLEANING RAILROAD CARS.

Section of Pipe.	Cubic Ft. Free Air per min.	Equivalent Length, feet.	Size of Pipe, In. Diam.	Average Vacuum Ins. Mercury.	Friction Loss, Ins. Mercury.	Final Vacuum, Ins. Mercury.
x—5	60	150	2½	6	0.35	6.35
5—4	120	140	2½	7	1.35	7.70
4—2	180	280	2½	11	7.0	14.70
2—w	180	190	3	16	4.0	18.70
w—u	360	20	5	19	0.9	19.60
u—s	480	20	6	20	0.5	20.10
s—sep	600	20	6	20	0.4	20.50

This loss will be the maximum that is possible under any condition as it is computed with three sweepers working on the three most remote inlets on laterals "xy" and "vw" and with two sweepers on laterals "tu" and "rs." The pipes are the largest which will give a velocity of 40 ft. per second with the full load and at the density which will actually exist in the pipe lines with the vacuum maintained at the separator of 20 in. mercury in all cases, except the pipe from "s" to separator. There the size was maintained at 6 in., as it was not considered advisable to increase this on account of the reduced velocity which would occur when less than the total number of sweepers might be working.

As bare floor brushes will be used for cleaning coaches it is not considered advisable to reduce the air quantity below that required by such renovators. However, when carpet renovators are used in Pullman cars and upholstery renovators are used on the cushions of both coaches and Pullmans, the air quantity will be reduced. This condition may exist at any time, also one of these carpet or upholstery renovators may be in use on one of the inlets most remote from the separator at the same time that nine floor brushes are in use on the remaining outlets. In that case a vacuum at the separator of less than 20 in. would result in a decrease in the vacuum at the inlet to which this renovator was attached. The vacuum at the separator must, therefore, be maintained at the point stated.

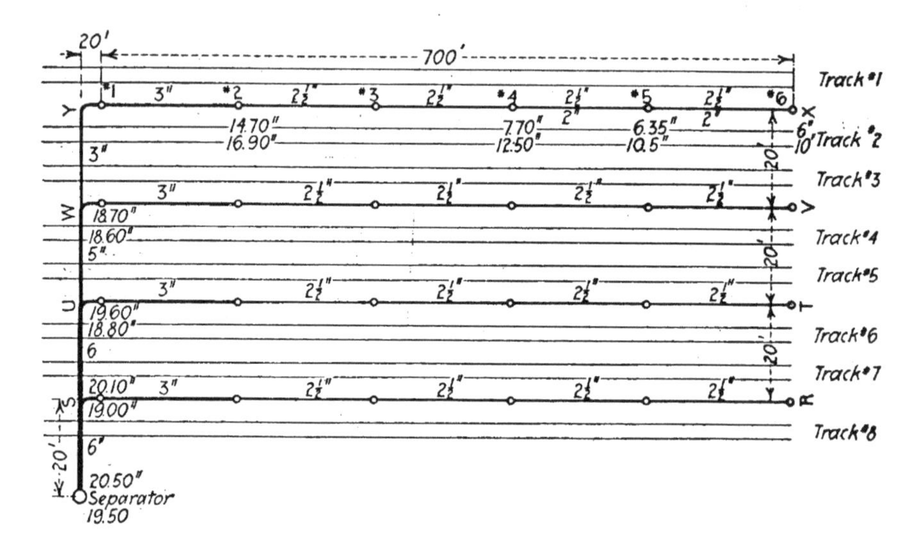

FIG. 63. VACUUM CLEANING LAYOUT FOR A PASSENGER CAR STORAGE YARD.

With such a vacuum there will be variation in the vacuum at the hose cocks of from 6 in. to 20 in. or seven times the maximum allowable variation in vacuum at the hose cocks.

If 1-in. hose be used, the maximum air quantities will be 40 cu. ft. per sweeper If we start with a vacuum at the inlet "x" of 10-in. mercury, the vacuum at the separator will again be 20 in. and we now have a variation of 10 in. between the nearest and most remote inlet from the separator, or five times the maximum allowed.

Either of these conditions is practically prohibitive, due to:

1. The excessive power consumption at the separator. 50 H. P. in case 1¼-in. hose is used, and 33 H. P. in case 1-in. hose is used.

2. The excessive capacity of the exhauster in order to handle the air at such low density, a displacement of 1,800 cu. ft. being necessary in case 1¼-in. hose is used and 1,200 cu. ft. in case 1-in hose is used.

3. The great variation in the vacuum at the hose cocks which will admit the passage of so much more air through a brush renovator on an outlet close to the separator as to render useless the calculations already made, or the high vacuum at the carpet or upholstery renovators would render their operation practically impossible.

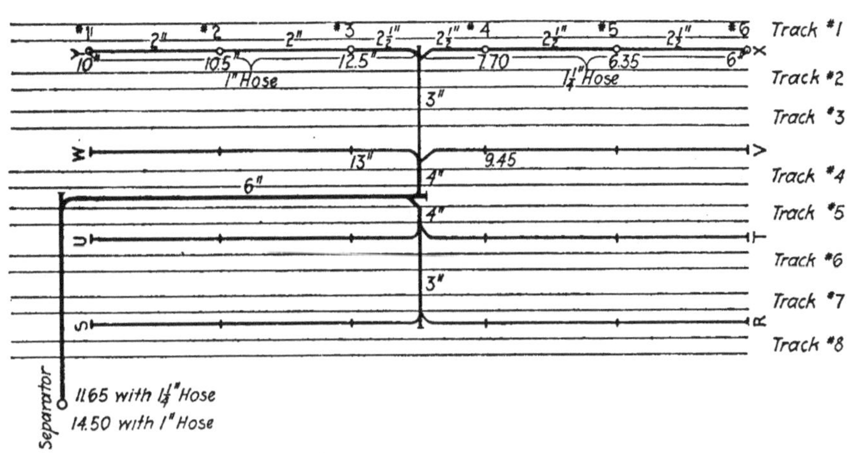

FIG. 64. ARRANGEMENT OF PIPING RECOMMENDED AS BEST FOR PASSENGER CAR STORAGE YARD.

Such a layout must be at once dismissed as impractical, and some other arrangement must be adapted. The arrangement of piping shown in Fig 64 is considered by the author to be the best that can be devised for this case.

With this arrangement the vacuum at the separator must be maintained at 11.50 in. mercury to insure a vacuum of 6 in. mercury at the outlet "x" under the most unfavorable conditions, and the maximum variation in vacuum at the inlets will be 3.45 in. mercury when 1¼-in. hose is used. This will give a maximum vacuum under a carpet renovator of 7½ in. mercury with 37 cu. ft. of air passing and will permit 70 cu. ft. of free air per minute to pass a brush renovator when operating with 100 ft. of hose attached to the inlet at which the highest vacuum is maintained. Both of these conditions will permit satisfactory operation and the increased air quantities will not seriously affect the calculations already made. The maximum horse power required at the separator will now be 20.5 as against over 50 in the case of the piping arrangement shown in Fig. 63, and will require an exhauster having a displacement of 950 cu. ft. instead of 1,800 cu. ft. required with the former layout.

If 1-in. hose is used and 10 in. mercury maintained at the outlet "x" under the same conditions as before, the vacuum at the separator will be 14.50 in. and the maximum variation in the vacuum at the inlets will be 3 in., which will give a maximum vacuum under a carpet renovator of 6 in. mercury with 32 cu. ft. of air passing and will permit the passage of 45 cu. ft. of free air through a brush renovator when operated at the end of 100 ft. of hose attached to the outlet at which the highest vacuum is maintained. This is a more uniform result, than was noted when 1¼-in. hose was used.

The maximum horse power which will be required at the separator will now be 18.6 and the maximum displacement in the exhauster will be 740 cu. ft.

It is, therefore, evident that, where very long runs of piping are necessary and where 100 ft. of hose will always be necessary, the use of 1-in. hose will require less power and a smaller displacement exhauster than would be required with 1¼-in. hose, without affecting the efficiency of the cleaning operations, and at the same time rendering the operation of the renovators on extreme ends of the system more uniform.

The example cited in Figs. 63 and 64 is not by any means an extreme case to be met in cleaning systems for car yards, and the larger the system the greater will be the economy obtained with 1-in. hose.

Such conditions, however, are confined almost entirely to layouts of this character and will seldom be met in layouts within any single building. This is fortunate, as the train cleaning is practically the only place where the use of 100 ft. of hose can be assured at all times.

Very tall buildings offer a similar condition although the laterals are now vertical and can be kept large enough to sufficiently reduce the friction without danger of deposit of dirt in them, and the horizontal branches will be short and also large enough to keep the friction within reasonable limits without danger of deposit of dust.

Where large areas within one or a group of buildings must be served by one cleaning system, better results can often be obtained by installing the dust separator at or near the center of the system of risers instead of close to the vacuum producer, as indicated in Fig. 65. When this is done, the pipe leading

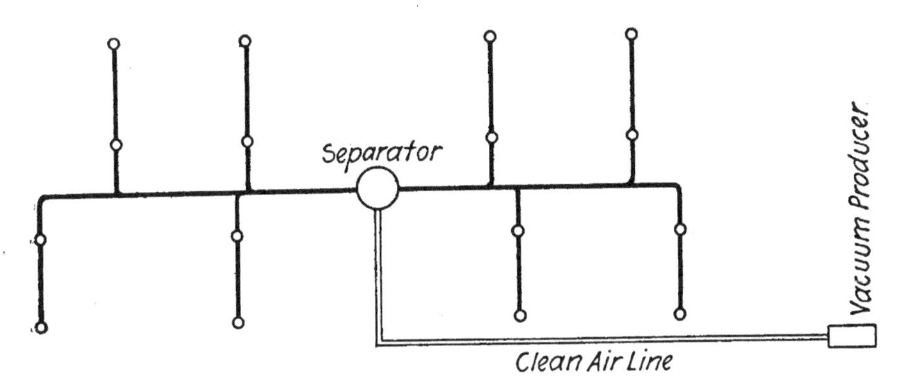

FIG. 65. GOOD LOCATION FOR DUST SEPARATOR WHERE LARGE AREAS ARE SERVED BY ONE CLEANING SYSTEM.

from the separator to the vacuum producer carries only clean air and can be made as large as desired and the friction loss reduced, resulting in a considerable reduction in the power required to operate the system.

Where the system becomes still larger, two or more separators located at centers of groups of risers can be used and clean air pipes of any desired size run to the vacuum producer (Fig. 66). When more than one separator is used care should be exercised in proportioning the pipe lines from the separators to the vacuum producer so as to have the friction loss from the vacuum producer to each separator the same in order to give uniform results at all inlets. This loss should also be kept as low as possible in order to prevent a high vacuum in a separator serving a portion of the system on which few sweepers are in operation. If low friction losses in the clean air pipe will require larger pipes than it is practical or economical to install, pressure reducing valves might be located in the clean air pipes near the separators to so regulate the vacuum at the separators and insure uniform results. A system of this kind might serve several premises and the air used by each be metered and the service sold much the same as heat and electricity. However, the power required to operate the system would be greater than that needed to operate a similar num-

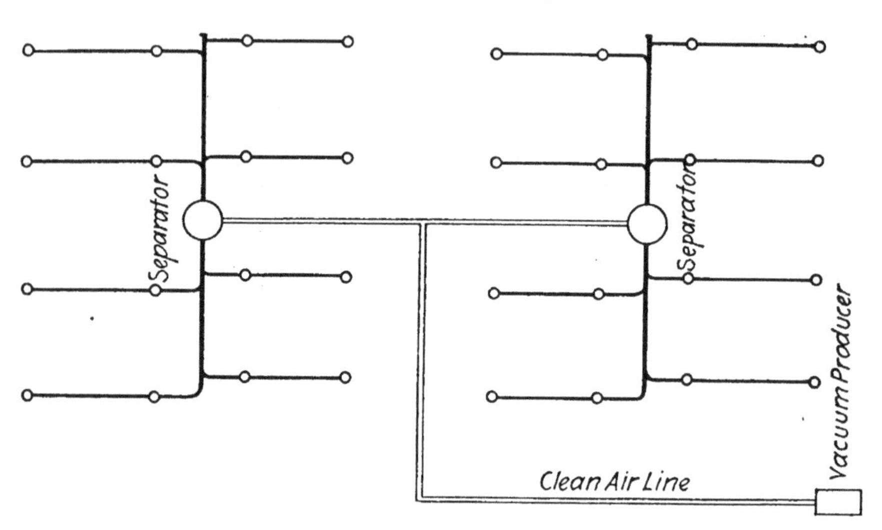

FIG. 66. LOCATION OF SEPARATORS AT CENTERS OF GROUPS OF RISERS FOR LARGE SYSTEMS.

ber of sweepers by individual plants owing to the higher vacuum required to overcome the friction in the trunk mains. This would be offset by the use of larger units and the possibility of operating them at full load at nearly all times. A system of this kind was contemplated in Milwaukee some seven years ago, but was never installed.

The question of pipe friction in connection with the design of vacuum cleaning systems requires careful consideration, much more than it ever received in the early days of the art and a great deal more than it sometimes receives at the present time.

CHAPTER VIII.

Separators.

The appliances which remove the dust from the air current which has carried it through the hose and pipe lines, in order to prevent damage to the vacuum producer, play an important part in the make-up of a vacuum cleaning system.

Classification of Separators.—Separators may be divided into two classes according to their use:

1. Partial separators, which must be used in conjunction with another separator in order to effect a complete removal of the dust from the air. These separators are again divided into two sub-classes, *i. e.,* primary, or those removing the heavy particles of dust and dirt only, and secondary, or those removing the finer particles of dirt which have passed through the primary separator.

2. Complete separators, or those in which the removal of both the heavy and the finer particles of dust is effected in a single separator.

Separators may also be classified, according to the method employed in effecting the separation, into dry separators in which all operations are effected without the use of liquid, and wet separators in which water is employed in the removal of the dust.

Primary Separators.—Primary separators are nearly always operated as dry separators and depend largely on centrifugal force to effect the separation. The first type of primary separator used by the Vacuum Cleaner Company is illustrated in Fig. 67. This consists of a cylindrical tank, with hopper bottom, containing an inner cylinder fixed to the top head. The dust-laden air enters the outer cylinder near the top on a tangent to the cylinder. The centrifugal action set up by the air striking the curved surface of the outer cylinder tends to keep the heavy dirt near the outside of same, and as it falls towards the bottom the velocity is reduced and its ability to carry the dust is lost. When the air passes below the inner cylinder the velocity is almost entirely destroyed and all but the very lightest of the dust particles fall to the bottom, while the air and the light dust particles find their way out of the separator through the opening in the center at the top.

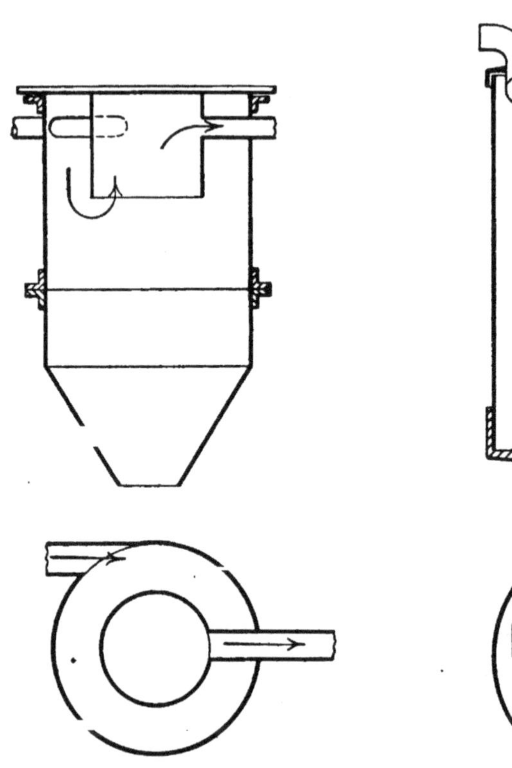

FIG. 67. EARLY TYPE OF PRIMARY SEPARATOR, USED BY VACUUM CLEANER COMPANY.

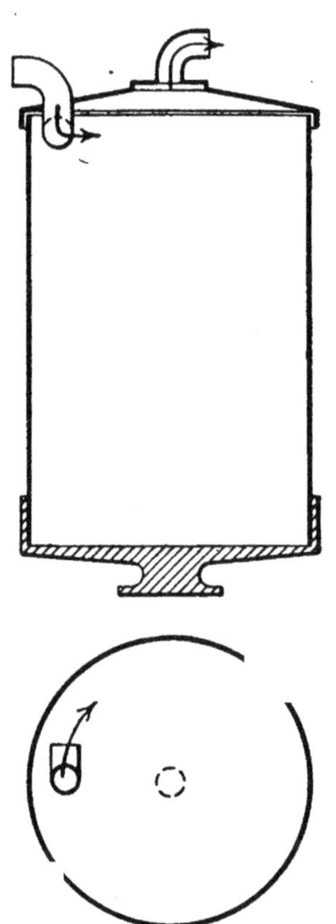

FIG. 68. PRIMARY SEPARATOR USED BY THE SANITARY DEVICES MANUFACTURING COMPANY.

The primary separator used by the Sanitary Devices Manufacturing Company is illustrated in Fig 68. The inner centrifugal cylinder is omitted and the air enters through an elbow in the top of the separator, near its outer extremity, which is turned at such an angle that the air is given a whirling motion resulting in the dust being separated much the same as in the case of the Vacuum Cleaner Company's apparatus.

Either of these separators will remove from 95% to 98% of the dirt that ordinarily comes to them through the pipe lines and are about equally efficient.

The separator illustrated in Fig. 69 was used by the General Compressed Air and Vacuum Cleaning Company. The entering air is led to the center near the bottom and is then released through two branches curved to give the air a whirling motion. The clean air is removed from the center of the separator near

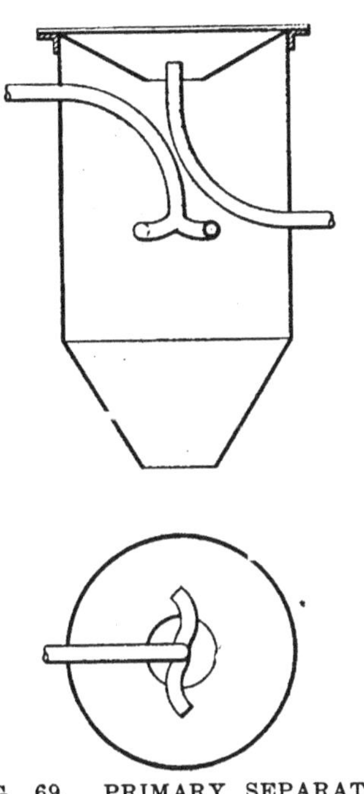

FIG. 69. PRIMARY SEPARATOR USED BY THE GENERAL COMPRESSED AIR AND VACUUM CLEANING COMPANY.

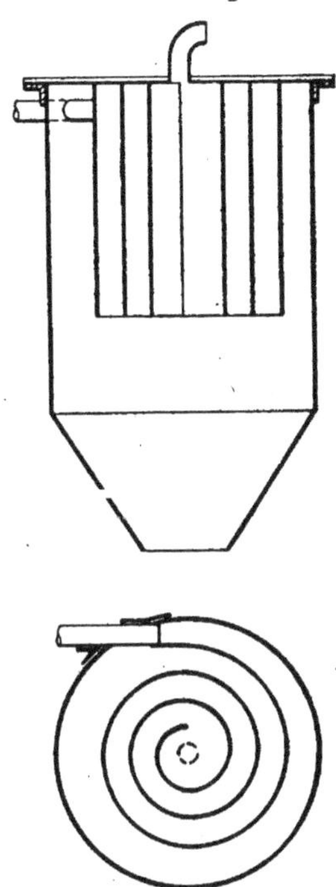

FIG. 70. PRIMARY SEPARATOR MADE BY THE BLAISDELL ENGINEERING COMPANY.

the top. This separator is not as effective in its removal of dirt as either of the former types, owing to the entering air being introduced near the bottom This tends to keep the air and the dust in the bottom of the separator continually stirred up, also the curved inlets give the air more of a radial than a tangential motion and there is less separation due to centrifugal action.

The separator illustrated in Fig. 70 is made by the Blaisdell Engineering Company. In this separator the inner centrifugal cylinder of the Vacuum Cleaner Company's separator is replaced by a spiral extending nearly to the outlet in the center of the top. This arrangement tends to prevent the reduction in the air velocity and to limit its effectiveness in the removal of dust.

Separators similar to the Sanitary separator have been manufactured by many firms producing vacuum cleaning systems. These all differ somewhat in details of construction but the principle involved, *i. e.*, centrifugal force and reduction in air velocity, is the same in all cases.

With vacuum producers in which there are no close clearances or rubbing contacts, these are the only separators used. The finer particles of dust passing these separators are carried harmlessly through the vacuum producer and through the exhaust to the outer atmosphere or to the chimney or other flue where they are effectively sterilized.

Secondary Separators.— With vacuum producers having close clearances or rubbing parts in contact with each other and the air exhausted, further separation of the finer dust particles is necessary. To accomplish this, secondary separators are used.

All of the early systems used a wet separator as a secondary separator. That used by the Vacuum Cleaner Company is illustrated in Fig. 71. It consists of a cylindrical tank partially filled with water, with a diaphram perforated in the central portion and fixed in place below the water line, and an inverted frustrum of a cone placed just above the water line. The air enters the separator below the water line and passes up through the water in the form of small bubbles which are broken up into still smaller bubbles on passing through the perforations in the diaphram. This action is very essential to the thorough cleansing of the air, as large bubbles of air may contain entrapped dust which will pass through the water and out into the vacuum producer. The inverted frustrum of a cone is intended to prevent any entrained water passing out of the separator with the air. This separator has always given satisfactory results when used in connection with reciprocating pumps.

The separator illustrated in Fig. 72 was manufactured by the General Compressed Air and Vacuum Cleaning Company. The air enters the separator through the pipe curved downward and escapes at the center below the water line. It then rises in the form of bubbles and most of it strikes the under side of the ribbed aluminum disc "a," which is intended to float on the surface of the water, and passes along the ribbed under surface of this disc, escaping into the upper part of the separator around the edge.

The clean air passes out of the top of the separator to the vacuum producer. The successful operation of this separator is

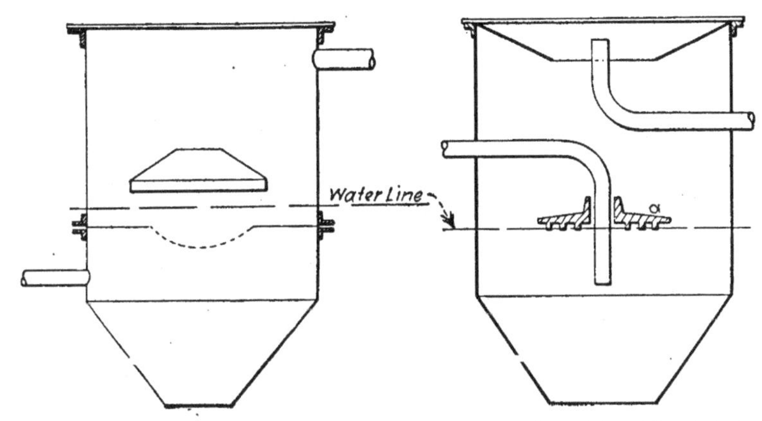

FIG. 71. SECONDARY SEPARATOR USED BY THE VACUUM CLEANER COMPANY.

FIG. 72. SECONDARY SEPARATOR USED BY THE GENERAL COMPRESSED AIR AND VACUUM CLEANING COMPANY.

dependent on the freedom of motion of the disc "a," which will always keep it on the surface of the water, and on all of the air passing up through the water under the disc.

Should the disc become caught on the supporting pipe the violent agitation of the water, which occurs when the system is in operation, will cause the disc to be left high and dry above the water at times, and submerged at other times. When this disc is above the water line it will not break up any of the large bubbles. Also, when there is a large quantity of air passing through the separator, there is great likelihood that considerable of the air bubbles will pass up through the water entirely outside of the disc and these bubbles will not be broken up. This separator has given somewhat unsatisfactory results in some installations tested by the author.

The separator used by the Sanitary Devices Manufacturing Company differs from those already described in that the air and water are mixed before they enter the separator and the air comes into the separator above the water line. The air enters the pipe "a" (Fig. 73) and passes to the aspirator "b," which is connected to the separator by the pipe "d" below the water line and the pipe "e" above the water line. The excess of vacuum in the separator draws the water out of the aspirator

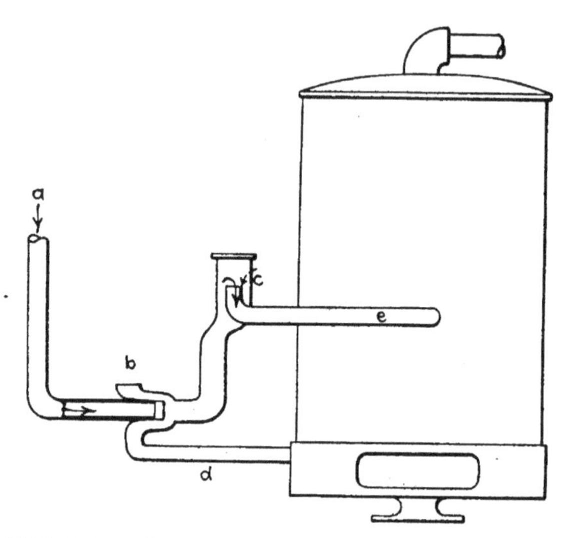

FIG. 73. SECONDARY SEPARATOR USED BY THE SANITARY DEVICES MANUFACTURING COMPANY.

and its pipe connections until the water line in this pipe is lowered below the top of the horizontal portion of the piping, when the air bubbles up through the demonstrator glass "c" and passes into the separator through the pipe "e." The filling of the vertical pipe leading to "c" with air causes the static head of the water in the separator to produce a flow of water through the pipe "d" into the aspirator "b." This is formed in the shape of a nozzle and the water enters in the form of a spray and thoroughly mixes with the air. The cleansing action of this water spray has been found to be very effective in removal of all fine dust and this separator has been found to be the most effective wet separator ever produced.

While the wet separator when properly designed will effectively remove the finest of dust, greasy soot will not emulsify with the water and its removal is practically impossible. Fortunately, this form of material in the finely-divided condition in which it passes the primary separator is not gritty and does not produce injurious effect on the vacuum producer.

The wet separator is also at a disadvantage in that there is a loss of vacuum in passing through same equal to the head of water that is carried between the inlet and the surface of the water. This generally amounts to nearly 2 in. mercury.

Means must be provided to observe the height of the water in the wet separator. For this purpose a glass window in the side of the separator has been found to be the most effective. The use of an ordinary gauge glass such as is used on boilers has been tried, but it has been found that they readily become so clouded by the action of the muddy water as to render them useless while the constant agitation of the water against the window when the system is in operation tends to keep the glass clean.

Dry separators have been used for secondary separators to a limited extent. All of these contained a bag made of canvas or some other fabric. The separator illustrated in Fig. 74 contains a bag made of drilling which is slightly smaller than the inner diameter of the cylindrical casing of the separator. The air enters the inside of the bag, inflating it, and passes through the bag and out through the opening in one side of the casing. A wire guard is placed over this opening to prevent the bag being drawn against the opening and thus rendering only a small portion of it effective.

These bags offer very little resistance to the passage of the air when they are clean but they soon become filled with dust and produce an increased resistance which, if neglected, may result in so great a difference in pressure as to hinder the action of the system and result in the rupture of the bag, letting the dust into the vacuum producer.

Some trouble has been experienced in finding a suitable material through which the dust will not pass. Hush cloth, such as used on dining tables, has been found to be the best material for this purpose. Better results are obtained by passing the air from the outside of the bag towards the inside than when the air is passed as indicated in Fig. 74. When this arrangement is adopted, it is necessary to stretch the bag over a metal screen or frame in order to prevent collapse.

Complete Separators.—Complete separators are of two classes, *i. e.*, dry and wet. The first complete separator that the author has knowledge of was used by the Vacuum Cleaner Company, in the form of a cylindrical tank and contained centrifugal cylinder and also a perforated plate. It was practically a combination of the separators indicated in Figs. 67 and 71. This separator was installed in connection with a small rotary pump

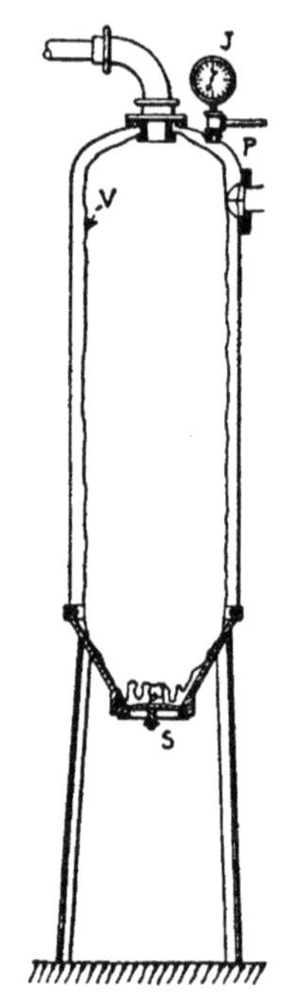

FIG. 74. TYPE OF DRY SEPARATOR USED AS SECONDARY SEPARATOR.

and mounted on a truck. It worked very well until it became filled with dirt when, in one case, the entire contents were ejected into an apartment in which it was being used. This separator was then rebuilt in the form shown in Fig. 75, the bag being made of hush cloth stretched over a wire screen. The air enters the cylinder tangentially and much of the separation is accomplished by centrifugal force, the remainder of the dust being removed as the air passes through the bag. This separator was successfully used as long as this company continued to manufacture such apparatus.

Another form of complete separator quite similar to that above described has recently been brought out by the Electric Renovator Manufacturing Company and is shown in section in Fig. 76. The air enters this separator tangentially below the line of the dust bag, which is made of muslin folded back and forward over a set of concentric cylinders thus giving a large area for the passage of the air. Being entirely above the line of the entering air, none of the heavy dirt strikes the bag and what dirt is caught on the bag is on the lower side of same and is shaken off every time the bag is agitated. This agitation

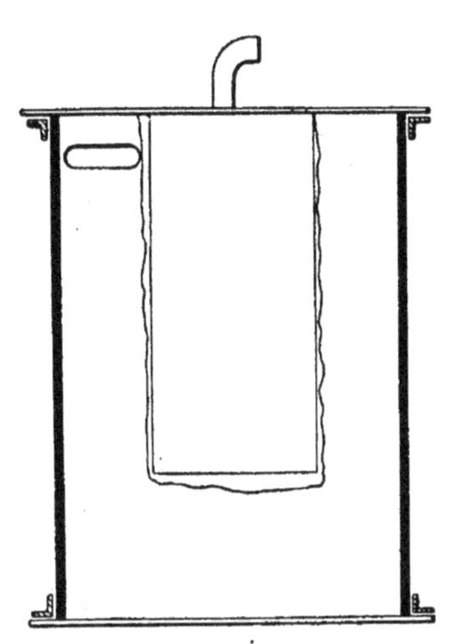

FIG. 75. FORM OF COMPLETE SEPARATOR USED BY THE VACUUM CLEANER COMPANY.

occurs every time there is any change in the volume of air passing the separator, and when these separators are used in connection with fan type of exhausters there is a constant surging whenever the exhauster is operated with a small volume of air passing. This tends to keep the bag clean automatically.

The separator illustrated in Fig. 77 is manufactured by the American Radiator Company. The air enters this apparatus through the pipe in the center and passes directly down to the bottom, the velocity being gradually reduced due to the expansion of the air as it passes down the cone-shaped inlet, the heavy dirt falling to the bottom. The air then passes up along

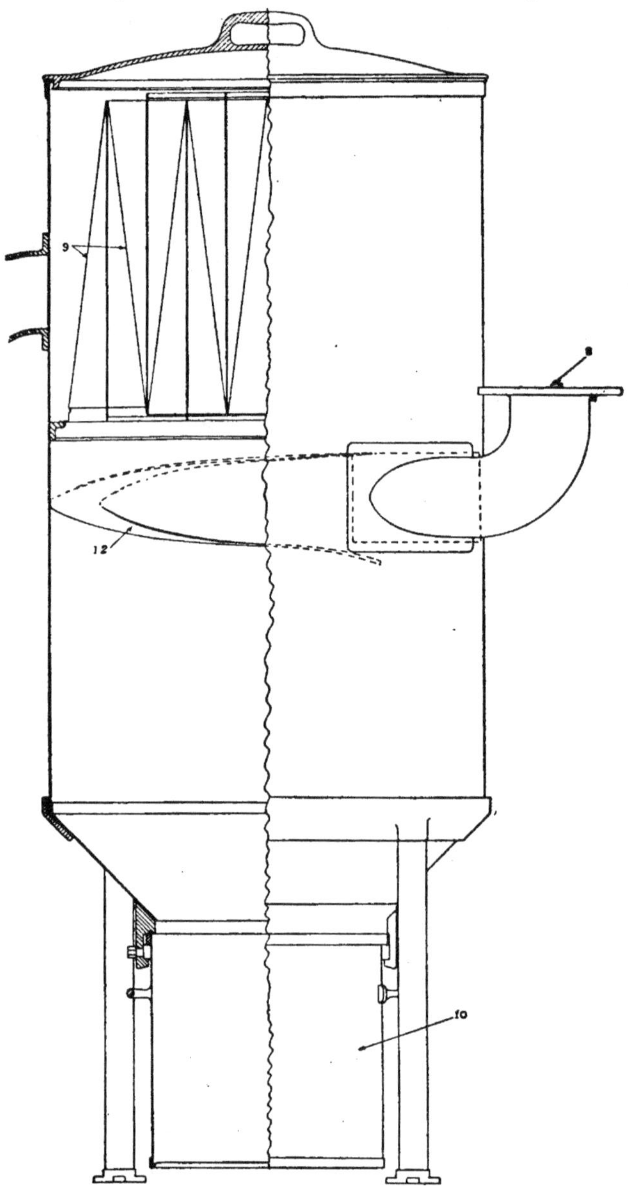

FIG. 76. COMPLETE SEPARATOR BROUGHT OUT BY THE ELECTRIC RENOVATOR MANUFACTURING COMPANY.

the inner surface of the cylindrical shell and thence through the bag, which is stretched over a screen, to the outlet. In this separator we see the first case in which centrifugal action is not utilized in separating the heavy dust, the makers evidently con-

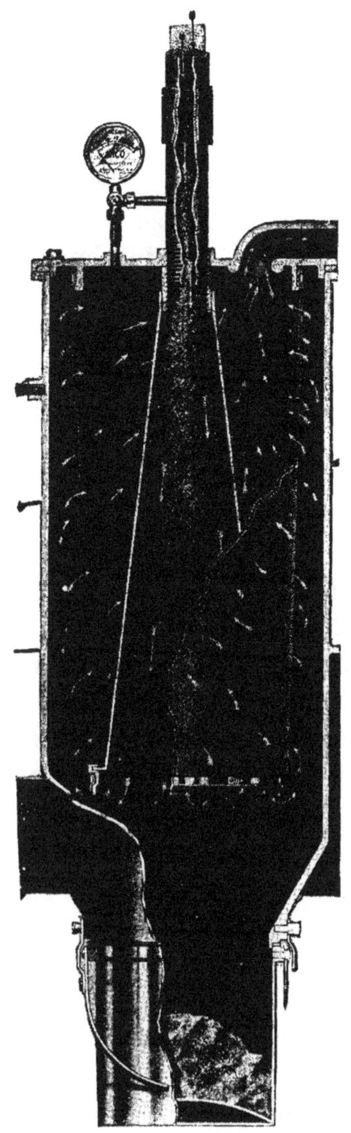

FIG. 77. COMPLETE SEPARATOR MADE BY THE AMERICAN RADIATOR COMPANY.

sidering the reduction of air velocity and the action of gravitation to be ample. This bag is arranged to permit the air passage from the outside towards the inside and it is tapered to allow the dirt to fall off. The vacuum gauge is connected to the inner and outer sides of the bag by means of a three-way cock to permit of measuring the difference in vacuum between the inside and outside of the bag to determine when the bag is in need of cleaning, which is accomplished by a reversal of the air current through the bag. This is quite necessary in order to keep the separator always in an efficient condition.

A separator was devised by the Sanitary Devices Manufacturing Company in which the bag was held extended by a wire ring having a weighted rod passing out through the top of the separator attached thereto. When the bag became clogged the difference in pressure on the two sides would result in a tendency of the bag to collapse and the rod would be raised up out of the separator, indicating that cleaning was necessary, which could be easily accomplished by drawing the rod up and down a few times thus shaking the dust off the bag. This separator never came into general use, although its arrangement was ingenious and should have been easy to operate.

The great difficulty with all bags which must be cleaned periodically is that they are almost universally neglected even when there is a visual indicator to show the accumulation of dirt, and when it becomes necessary to manipulate a three-way cock in order to ascertain when this cleaning must be done it will seldom if ever be attended to. A bag that will clean itself, such as the Capitol Invincible, is shown in Fig. 76.

The separator used by one manufacturer consists of a simple cylindrical tank into which the air is blown tangentially, with a screen near the top, the whole forming a base for the vacuum producer. This separator does not remove any but the heaviest dirt and is suitable for use only with a vacuum producer having very large clearances and in locations where the discharge of considerable dirt into the atmosphere is not objectionable.

Total Wet Separator.—The only total wet separator which is in commercial use is manufactured by the American Rotary Valve Company. This separator is contained in the base of the vacuum producer and is provided with a screen near the point of entrance of the dust-laden air, which screen is cleaned by a mechanically-driven bristle brush. When the water in the separator becomes foul, the contents of the separator are discharged direct to the sewer by means of compressed air. If this separator receives proper attention it makes the most sanitary arrangement that has been introduced in the vacuum cleaning line to date. However, the separator should be emptied at frequent intervals or the volume of solid matter contained in the same will become so great that there will not be enough water present to flush the sewer and stoppage is likely. These separators are often neglected until the contents become of the consistency of mortar or molasses which is not a fit substance to discharge into a sewerage system.

There is still another form of apparatus used in connection with vacuum cleaning systems which should be called an emulsifier rather than a separator. That is the type used with the Rotrex and the Palm systems. The dust is mixed with water when it first enters the pump chamber, a screen being used to remove the lint and larger particles of dirt and then the mud produced by the combination of the dust and water is passed through the pump along with the air. The air and muddy water are separated on the discharge side of the vacuum producer. In many cases where the exhaust pipe is long, there is considerable back pressure on the discharge which is often sufficient to force the seals in traps on the sewerage system, allowing sewer gas to be discharged into the building in which the cleaning system is installed. No means are provided for automatically cleaning the screen used in these appliances and the author knows of cases where the screen has become so completely clogged with lint that its removal from the machine was necessary in order to render the operation of the cleaning tools possible.

When dry separators are used, the manual removal of the dry dirt accumulated is necessary and is an objectionable as well as unsanitary operation. The author considers that the ideal arrangement of separator would be one in which the dirt can all be emulsified with water and retained in the separator, only the air passing through the vacuum producer, and in which the contents of the separator would be discharged automatically to the sewer when the density of this mixture becomes as heavy as will readily run through the sewer. This discharge should be of sufficient volume to completely fill an ordinary house sewer in order to insure a thorough flushing of the drain, and should be discharged into the sewer under atmospheric pressure in order to guard against the forcing of water seals in any of the plumbing fixtures.

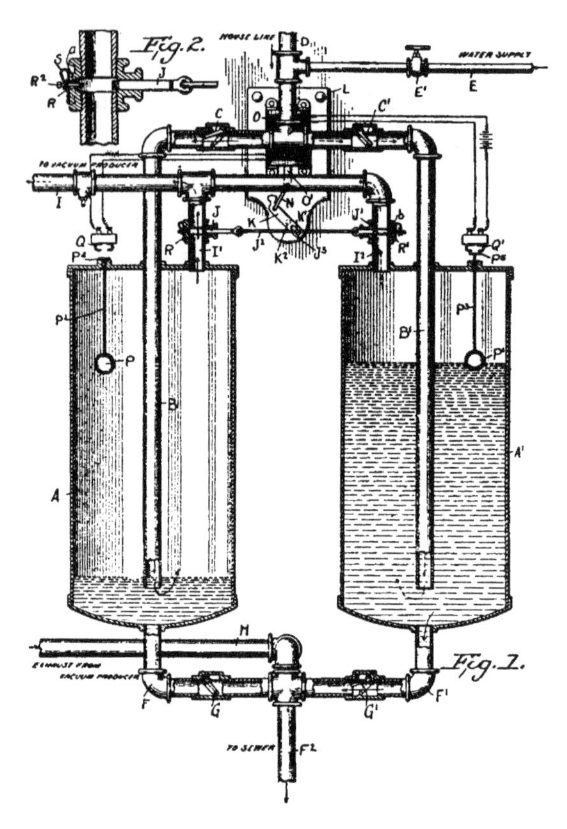

FIG. 77a. INTERIOR CONSTRUCTION OF DUNN VACUUM CLEANING MACHINE.

A separator of this type has recently been patented by E. D. Dunn, originator of the Dunn Locke system. It is illustrated in Fig. 77a. The action of the separator is as follows: After starting the motor and turning on a small quantity of water, a vacuum is produced in one tank and through a system of piping to the cleaning implement in use. The dust and dirt collected by the implement is saturated as it approaches the plant and in this saturated condition enters the bottom of a body of water in the tank.

When the accumulating dirt and water reach a certain level a valve is automatically operated which closes the tank's communication with the vacuum pump and allows its contents to flow off to the sewer by gravity. The mechanism for operating the valve is rather unique and includes a float which, on rising with the water, makes a positive electrical contact, as shown in the figure. In this illustration one tank is about to discharge and the other tank is about to become operative. The electrical contact causes the core of the magnet at 0′ to rise, making the lever, K, turn over, which action opens one valve and closes the other. In this way the tanks alternately partly fill and empty their collections of water and sweepings.

This system has not as yet been in commercial use for a sufficient length of time to insure its successful operation, and the author does not consider the passing of dirt and water through ordinary check valves to be commercially possible without rendering these checks inoperative.

Check valves have been used where partial wet and dry separators are operated in tandem to prevent drawing water into the dry separator, in the event of the plant being shut down with all inlets on the pipe line closed. In such a case, the leakage through the pump into the wet separator may raise the pressure in this separator faster than leakage on the pipe line raises the pressure in the dry separator.

This is accomplished by providing a small connection between the upper part of the two separators, fitted with a check valve opening towards the dry separator. When the vacuum producer is in operation, the vacuum in the wet separator is approximately 2 in. greater than that in the dry and the check is held closed. When the vacuum producer is stopped and the vacuum in the wet separator falls faster than in the dry separator, this check opens and clean air passes from the wet to the dry separator. When operating under these conditions, the action of the check valve is satisfactory. However, the author has known of cases where the check leaked and when this happened the check was immediately clogged by the dust-laden air from the dry separator.

CHAPTER IX.

Vacuum Producers.

The next portion of the cleaning system is that which produces the motion of the air through the system and that to which the motive power is applied, namely, the vacuum producer.

Types of Vacuum Producers.—Vacuum producers can be divided into general classes: 1. Displacement type, in which a constant volume of air is displaced during each complete cycle of operations of the machine, and 2. Centrifugal type, in which the volume of air passing the producer during each complete cycle of operations varies with the resistance to the passage of such air through the system.

Displacement Type.—Under this head the piston and rotary pumps are classed, and they are subdivided according to construction into reciprocating and rotary, valved and valveless, air cooled and water cooled.

Centrifugal Type.—Under this head the fan type of vacuum producers are classed. They may be divided, according to construction, into single stage and multi-stage, horizontal and vertical.

Power Required to Produce Vacuum.—In order to ascertain the efficiency of the various types of exhausters to be discussed in this chapter it is necessary to ascertain the actual power necessary to move one cubic foot of free air at any degree of vacuum.

As nearly all machines tested by the author were driven by electric motors and the power was, therefore, indicated in watts, the curve C-D in Fig. 78 showing the actual power necessary to exhaust one cubic foot of free air at the vacuum noted in the lower margin, assuming no clearance and adiabatic compression, is used as a basis for calculation of efficiency. This shows that to produce a vacuum of 8 in. mercury there will be required an expenditure of 16 watts for each cubic foot of free air exhausteed, and to produce a vacuum of 12 in. mercury will require an expenditure of 27 watts. If these quantities be divided by the efficiency of the machine the actual power required will be determined.

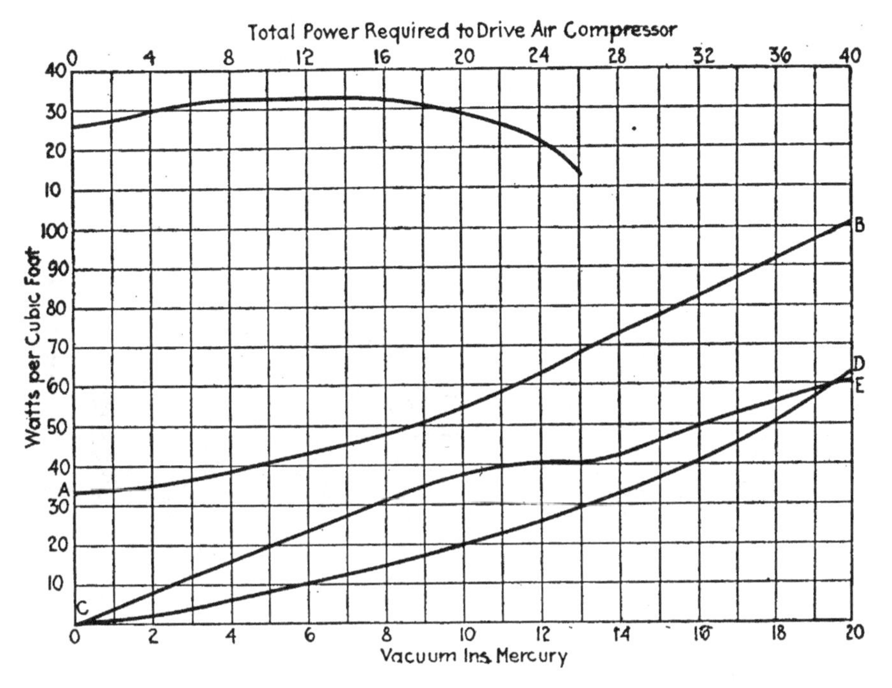

FIG. 78. POWER CONSUMPTION AND EFFICIENCY OF AIR COMPRESSOR USED AS A VACUUM PUMP.

Reciprocating Pumps.—The reciprocating pump was used on the majority of the earlier vacuum cleaning systems. The most common form in early use was a commercial air compressor which was used as a vacuum pump without any change in its construction. It was usually fitted with mechanically-operated induction and poppet type of eduction valves of heavy pattern, fitted with cushions of the dash pot principle, the same as are used on air compressors working against terminal pressures as high as 100 lbs. per square inch. The cylinders were water jacketed to remove the heat of high compression. The valves in these compressors were heavy and required considerable pressure to open them and the friction of the valve gear and other moving parts, which were made heavy enough to withstand the strains of high compression, was excessively high for a machine where the compression did not exceed 8 or 9 lbs. per square inch. Their efficiency, therefore, is lower under actual operating conditions than if they were working against pressures for which they were designed. A curve of the power consumption of a 14-in. x 8-in. Clayton compressor is shown on Fig. 78, the abscissae being the vacuum in inches of mercury and the ordinates of curve "AB" the watts required to exhaust one cubic

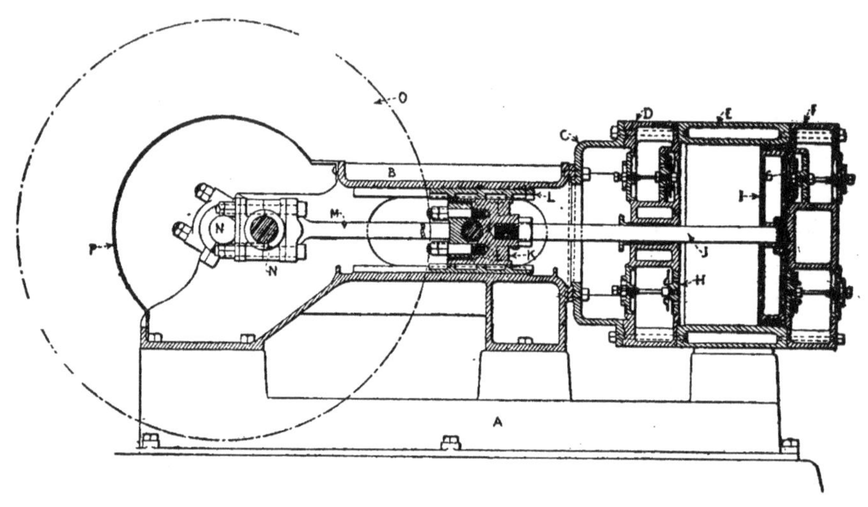

FIG. 79. MODIFICATION OF RECIPROCATING PUMP MADE BY THE SANITARY DEVICES MANUFACTURING COMPANY.

foot of free air. Curve "cd" represents the theoretical watts required to do the same work. These compressors were used in connection with systems operating with 1-in. hose and the vacuum usually carried was 15 in. mercury. They require approximately 77 watts per cubic foot of free air at this vacuum and the efficiency, shown in curve "ce" (Fig. 78) is 46%.

Were this compressor used in connection with a system operating through 1¼-in. hose and a vacuum of 8 in. mercury maintained, the efficiency would drop to 31%.

A modification of the reciprocating pump was manufactured by the Sanitary Devices Manufacturing Company in which light-weight poppet valves placed in the heads of the cylinder were used, as indicated in Fig. 79. Curves of the watts per cubic foot and efficiency of this type of compressor are shown in Fig. 80. It will be noted that this compressor shows a better efficiency than the air compressor at all degrees of vacuum and it is the best reciprocating pump that the writer has ever tested.

This pump was made for several years without water jacket and no trouble was ever experienced with overheating. However, owing to the commercial air compressors being jacketed, the makers using same made this a talking point and this company was obliged to jacket its pumps.

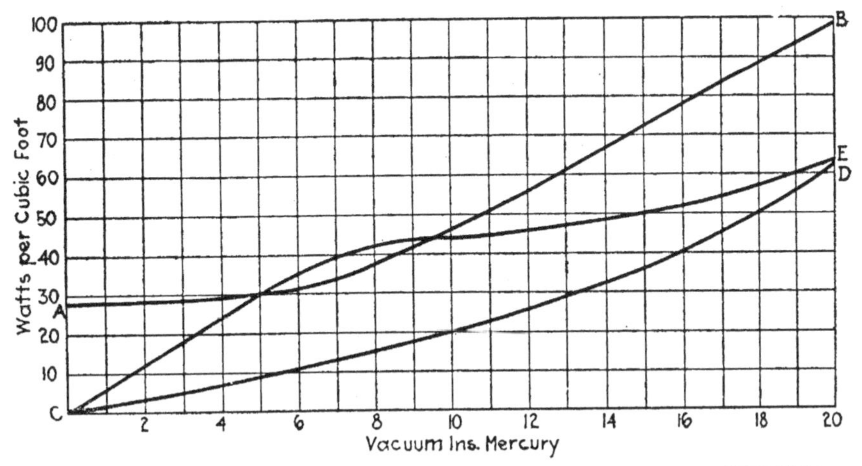

FIG. 80. POWER CONSUMPTION AND EFFICIENCY OF MODIFIED RECIPROCATING PUMP.

The Vacuum Cleaner Company used a Clayton pump on its smaller plants which was fitted with a semi-rotary valve in each end serving as an induction and eduction valve, while the heavy poppet eduction valve of the air compressor was dispensed with. The increase in efficiency that should have resulted from this change was not realized. The reason for this can be more readily seen by inspection of the indicator cards, Figs. 81 and 82.

Fig. 81 is a card taken from one of the Clayton compressors fitted with combined induction and eduction valves, and Fig. 82 a card from a compressor with light steel induction and eduction valves of the poppet type.

It will be noted that the compression line, a-d, Fig. 81, extends above the atmosphere line, the pressure at the time of opening the eduction valve being 4 lbs. per square inch above the atmosphere. This is due to the failure of the mechanically-operated valve to open soon enough. This valve being also the induction valve, it is necessary for the eduction port to be closed before the induction port can be opened, in order to prevent a short circuit of air from the atmosphere into the separators. This fact is responsible for the sudden increase in the pressure at b, the eduction port having closed before the completion of the stroke and the air in the clearance space being compressed

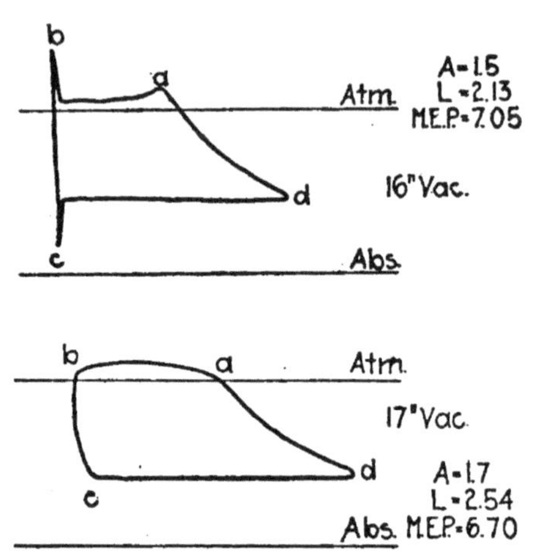

FIGS. 81 AND 82. INDICATOR CARDS FOR CLAYTON AND MODIFIED PUMPS.

to 6½ lbs. above atmosphere. The induction port is not opened until after the beginning of the suction stroke resulting in the high degree of vacuum at c.

Compare this with the card, Fig. 82. Here the compression does not extend above the atmosphere line more than ¼ lb. per square inch and the eduction valve does not close until the end of the stroke so that the vacuum at the beginning of the suction stroke is no lower than during the entire stroke.

These pumps were working under the same conditions, *i. e.*, 15-in. vacuum in the separator. The M. E. P. for Fig. 81 is 7.05 while that in Fig. 82 is 6.7 and is higher than is usually the case with this pump, due to the fact that the exhaust pipe from this pump was very long and crooked, a condition which should be avoided whenever possible. Also, the pump from which this card was taken is one of the older pattern and the clearance was greater than in the later models. The point at which the eduction valve opens in Fig. 81 is 53% of the stroke and it closes at 95% of the stroke and is, therefore, open 42% of the stroke, while in Fig. 82 the eduction valve opens at 46% of the stroke and remains open to the end of the stroke, and, therefore, is open for 54% of the stroke. Thus the pump with the poppet valves will move more air at the same vacuum with less expenditure of power than the pump with the mechanically-operated valves.

Another type of reciprocating pump has been introduced in the past two or three years in which a single valve which rotates continuously in one direction is used for induction and eduction valve, for both ends of the cylinder. This valve is a plain cylindrical casting having ports cored through to alternately connect the cylinder ports with the intake and exhaust ports.

By rotating this valve 180° on its stem the vacuum pump is changed to an air compressor. This arrangement is adopted in order to discharge the contents of the separator into the sewer as was explained in Chapters I and VIII. In this pump there must be points at which both the induction and eduction valves are closed at the same time and results similar to those found with the semi-rotary valves of the Clayton pump will naturally be in evidence. The author has endeavored to obtain an indicator card from one of these pumps but has been unable to do so. The effect of simultaneous closing of both induction and eduction ports would naturally be more marked in this pump than in the Clayton, as the motion of the valve in this case is uniform at all times while the motion of the valve gear of the Clayton pump is so arranged that the valve moves very fast at the time that both ports are closed. One of the two pumps of this type which was recently installed in the New York Post Office is illustrated in Fig. 83. These pumps have a displacement of 1,200 cu. ft. each and are the largest reciprocating pumps in use for vacuum cleaning at this writing.

An interesting property of the piston pump which lends itself to the economical control of the vacuum in the system is illustrated by the curve at the top of Fig. 78 which shows the total power required to operate the Clayton type air compressor, the efficiency of which is indicated by the lower curves on this figure. The compressor was operated at constant speed and the air volume varied to give various degrees of vacuum from atmospheric pressure to a closed suction and the power to operate the compressor read at intervals of two inches. The current input to the motor in amperes is indicated by ordinates and the vacuum in the separator by the abscissae. This indicates that the piston pump requires the maximum power to

FIG. 83. ONE OF THE PUMPS INSTALLED IN CONNECTION WITH THE VACUUM CLEANING SYSTEM IN THE NEW YORK POST OFFICE, THE LARGEST RECIPROCATING PUMP USED FOR THIS PURPOSE UP TO THE PRESENT.

operate at about 15-in. vacuum and that the least power is required when the vacuum is at the highest point possible to obtain. The method employed in utilizing this characteristic of a piston pump will be discussed in a later chapter.

Rotary Pumps.—The Garden City rotary pump is a good example of the single-impeller type of pump and is or has been used to some extent by at least two makers of vacuum cleaning systems. Its interior arrangement is shown in Fig. 84. A solid cylindrical impeller, A, is mounted eccentrically in the cylindrical outer casing, the impeller being fitted with four sliding vanes which are provided with distance pieces, E, and wearing faces, B. The oil reservoir is provided with a needle valve which is automatically opened as soon as there is any vacuum produced and closes automatically when the machine is shut down. The rate of feed of oil is adjusted by the screw I. This type of pump offers a large surface in rubbing contact with the case and becomes very hot when in operation. It requires liberal lubrication in order to prevent heating and cutting of the surface of the casing. End wear in these pumps causes leakage, and, as usually constructed, there are no means

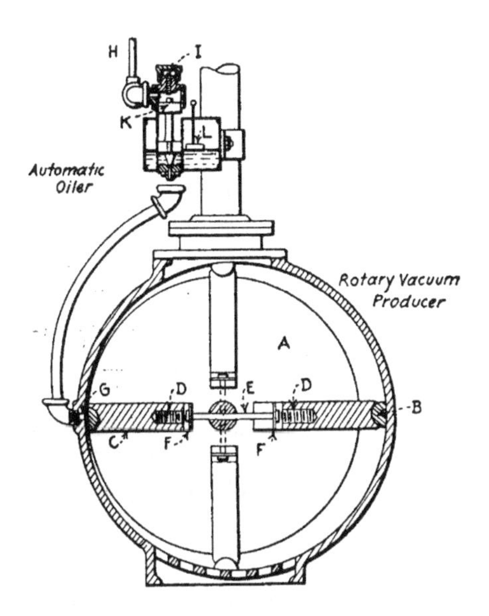

FIG. 84. INTERIOR ARRANGEMENT OF THE GARDEN CITY ROTARY PUMP.

provided for taking up this wear. It can be provided for, however, by using metal shins on the ends of the cylindrical casing.

The power required to operate this type of pump (Curve a-b, Fig. 85), is nearly the same as that required to operate a piston pump for vacuum less than 12 in. mercury, but when the vacuum becomes higher, the power required becomes much greater than that required by the piston pump. The efficiency (Curve c-e, Fig. 85), is identical with that obtained with the light-weight poppet valve pump (Curve c-e, Fig. 80) from 0 to 11 in. vacuum, but for higher vacuum the efficiency of this type of pump falls off, while the efficiency of the piston pump becomes greater as the vacuum becomes higher. This difference in the characteristics of the two types of pumps is due to the presence of valves in one case and their absence in the other. With the piston pump the atmospheric pressure reaches the cylinder only while air is being discharged, the eduction valves being closed at other times and a partial vacuum exists on both sides of the piston. The higher the vacuum produced, the less time there is atmospheric pressure on the piston until, when no air

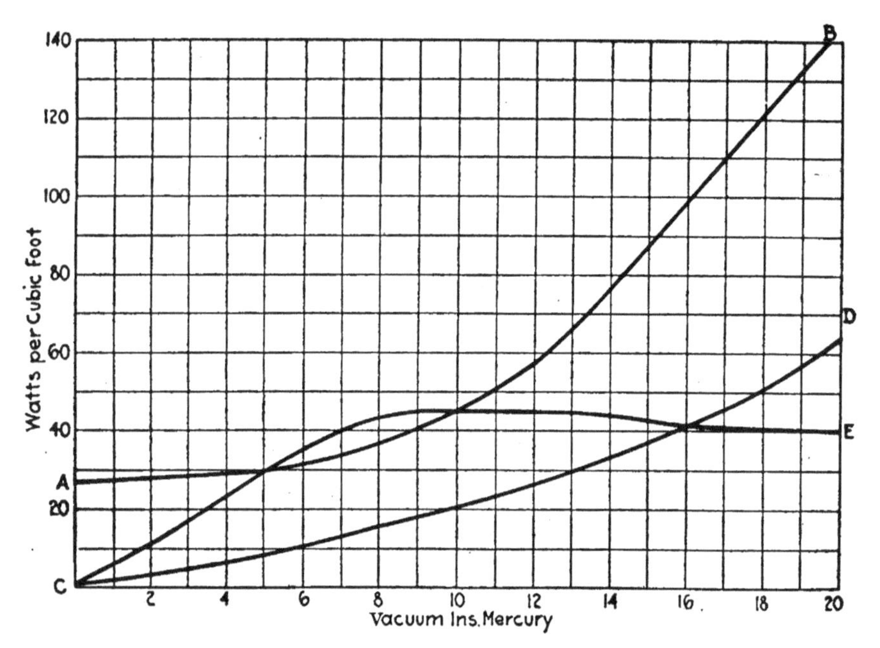

FIG. 85. POWER REQUIRED TO OPERATE GARDEN CITY TYPE OF ROTARY PUMP.

is discharged, the air contained in the clearance space of the cylinder is compressed and expanded, the compression and expansion lines being coincident. The indicator card will have no area, and the only power expended is that required to overcome the friction in the moving parts. With the rotary pump there are no discharge valves to hold the atmospheric pressure from the discharge side of the impeller and the compression of the rarified air is accomplished by the atmospheric pressure admitting air through the eduction port into the chamber. As it comes opposite the eduction port there is no difference in the time during which the impeller is subject to atmospheric pressure, no matter what the quantity of air being discharged. The higher the vacuum in the spaces containing rarified air, the greater the difference in pressure on the opposite sides of the sliding vane and, therefore, the greater total power required to turn the rotor.

FIG. 86. ARRANGEMENT OF DOUBLE-IMPELLER ROOT TYPE ROTARY PUMP FOR VACUUM CLEANING WORK.

Another type of rotary pump which is fast becoming the most popular is the double-impeller type. This is generally known as the Root blower, as the firm of this name was the first to manufacture same. They have been in use for many years as blowers for gas works, and as vacuum producers for various purposes, mainly the operation of pneumatic tube systems.

Why this form of vacuum producer was not earlier adopted in vacuum cleaning systems, instead of the sliding-vane type, is hard to understand. This pump contains two impellers or cams which are mounted on shafts geared together and revolve in opposite directions inside of a case, always being in close proximity to the case and to each other, but never touching. They are, therefore, frictionless in operation and the introduction of a small amount of water renders them practically air tight. There being no metallic contact between the moving parts, internal lubrication is unnecessary and there is no wear on either the impellers or the casing and no means of taking up wear are necessary.

The arrangement of the impellers and the method of providing water to seal the parts is shown in Fig. 86. A reservoir containing water is provided on the discharge side of the pump and a small pipe leads from this reservoir to the suction side of the pump. The vacuum lifts water from the reservoir and discharges same in a spray into the suction chamber. This water passes through the pump and is separated from the air in the discharge chamber to be returned to the suction chamber by the

FIG. 87. ROTARY PUMP ARRANGED WITH DOUBLE-THROW SWITCH FOR REVERSING PUMP.

vacuum. This operation will start automatically as soon as any degree of vacuum is formed and will cease as soon as the pump is shut down.

Any of these rotary pumps having no valves can be changed to an air compressor by reversing the direction of rotation. This is adapted by the American Rotary Valve Company in connection with their wet separators to discharge the contents of the separator into the sewer, on all of their smaller-sized plants. Fig. 87 shows one of these plants arranged with double-throw switch for reversing the electric motor used to operate the pump and also shows the arrangement of the rotary brush which is used to clean the screen in the wet separator, as has been explained in Chapter VIII.

The power consumption and efficiency of this type of pump are shown in Fig. 88. The watts per cubic foot of free air

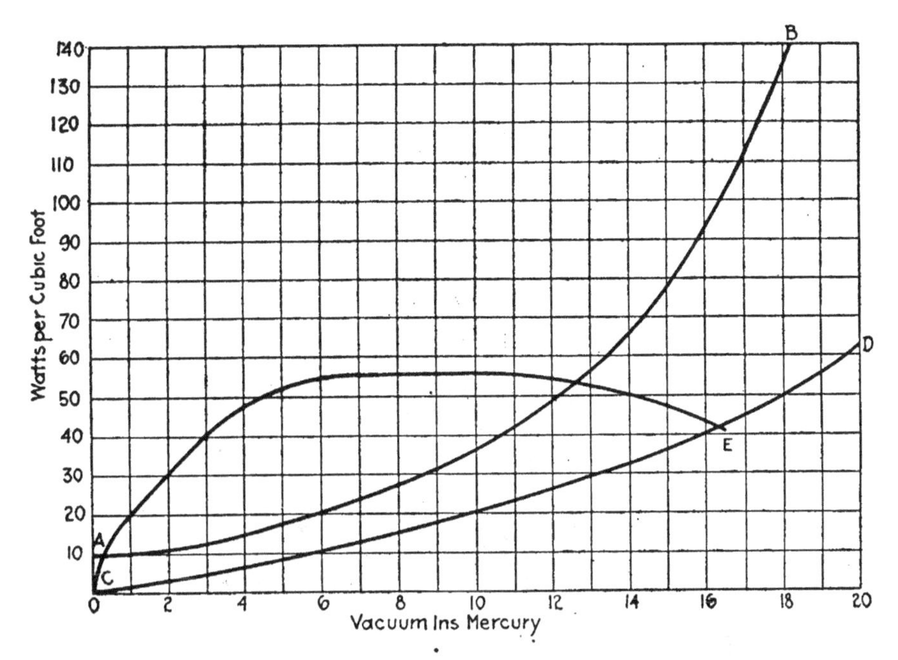

FIG. 88. POWER CONSUMPTION AND EFFICIENCY ROOT TYPE OF PUMP.

(Curve a-b) show a much lower consumption of power at the lower vacuum than any of the pumps already tested. This is probably due to the fact there is no internal friction. It will

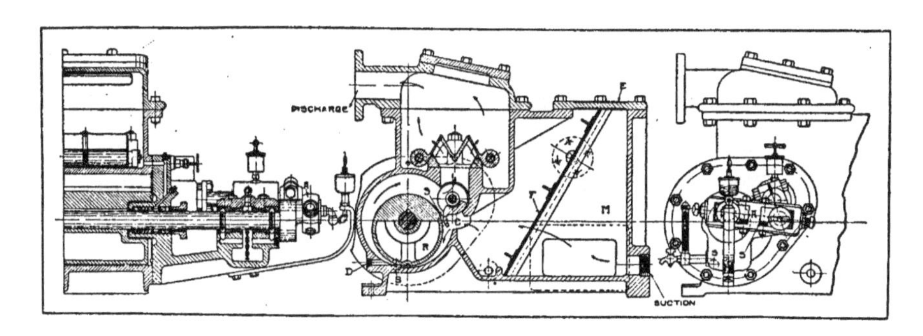

FIG. 89. THE ROTREX VACUUM PUMP, USED BY THE VACUUM ENGINEERING COMPANY.

be noted that the power to operate at no vacuum is but 10 watts per cubic foot of free air, while all the others require from 24 to 34 cubic feet. This also results in the efficiency curve (c-e, Fig. 88) reaching its maximum value at a lower vacuum than in the case of the sliding vane pump (Fig. 85).

The efficiency is fairly constant between 6-in. and 10-in.

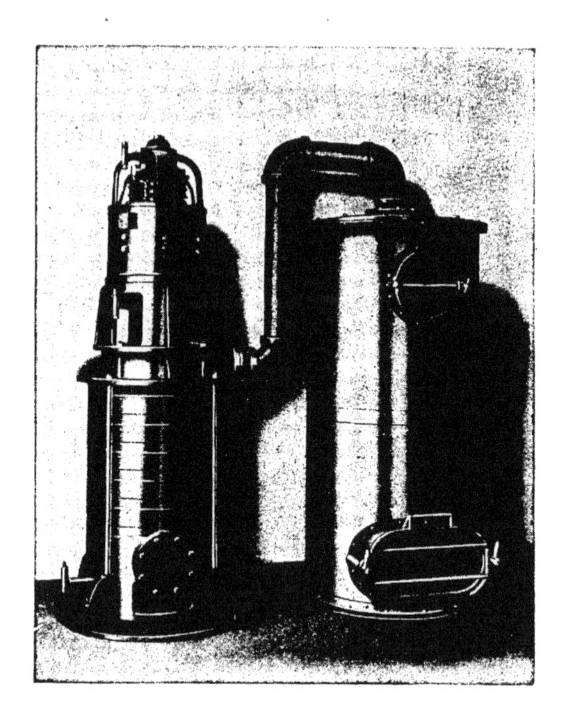

FIG. 90. LATE TYPE OF CENTRIFUGAL EXHAUSTER MADE BY THE SPENCER TURBINE CLEANER COMPANY.

vacuum and is much higher than is obtained with any of the other types of pumps at these vacua. When they are operated at higher vacuum the efficiency is about the same as obtained with the sliding vane pumps and lower than that obtained with the reciprocating pumps. The best efficiency of this pump is at the vacuum necessary to operate a cleaning system provided with 1¼-in. hose.

A slight modification of this type of pump is that used by the Vacuum Engineering Company, known as the Rotrex. This pump has but one impeller, of nearly the same form as the impellers in the Root blowers and has a follower driven by crank and connecting rods which is always in close proximity to the impeller but does not touch same. The arrangement of

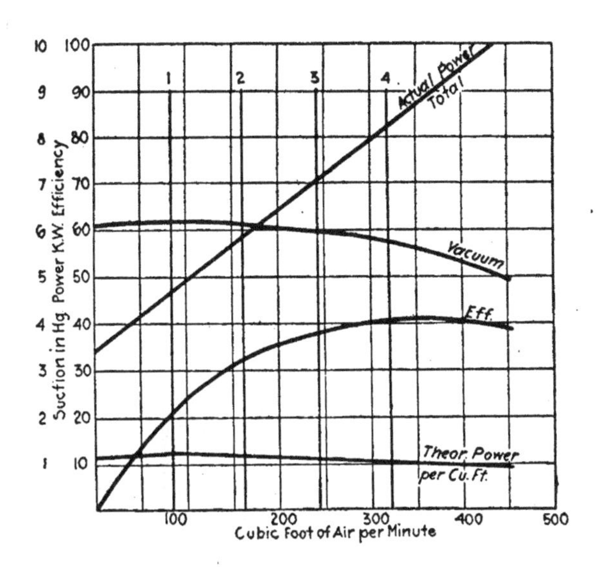

FIG. 91. POWER AND EFFICIENCY CURVES FOR THE SPENCER MACHINE.

this pump is illustrated in Fig. 89 which also shows the saturation chamber and screens used instead of a separator, as explained in Chapter VIII.

The author has never tested the economy of these pumps but would infer that their economy should be about the same as that of the Root blower.

Centrifugal Exhausters.—This type of exhauster has always taken the form of a fan. The first stationary fan type of exhauster was manufactured by the Spencer Turbine Cleaner Company. Their latest type is illustrated in Fig. 90. It consists of a series of centrifugal fans mounted on a vertical shaft, stationary deflection blades being provided between the wheels to conduct the air from the periphery of one wheel to the center of the next.

These centrifugal exhausters do not have a positive displacement, as do all those already described, and therefore the variation of the vacuum is not as much as in case of the positive displacement machines. The vacuum produced when the machine is moving no air is slightly less than the maximum that

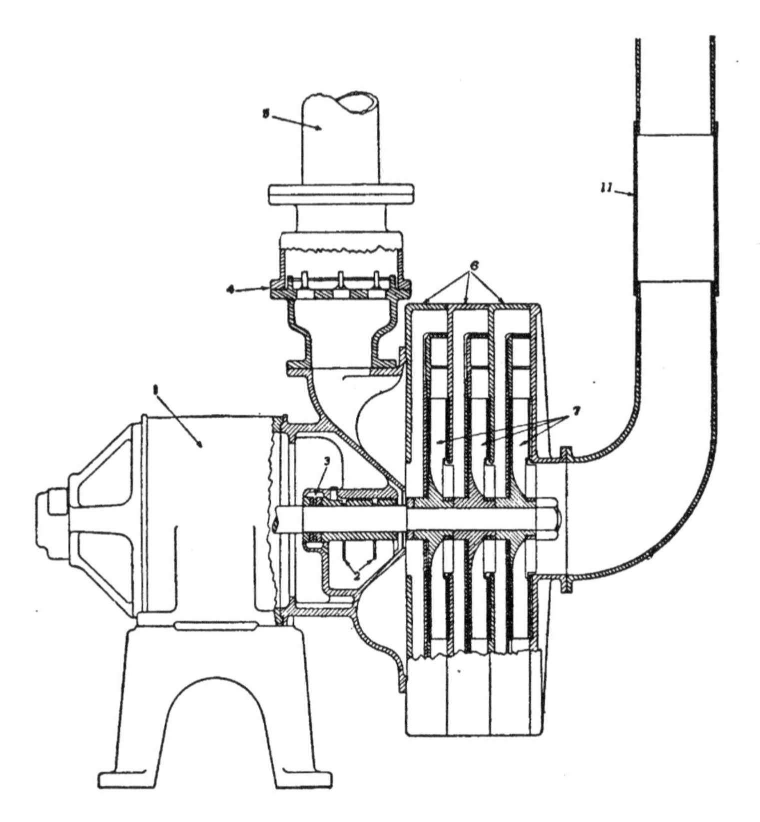

FIG. 92. INTERIOR ARRANGEMENT OF INVINCIBLE MACHINE, MANUFACTURED BY THE ELECTRIC RENOVATOR MANUFACTURING COMPANY.

the exhauster can produce and there is very little variation in the vacuum with air quantities which can be moved without exceeding the capacity of the motor or other means producing the power. The curves showing the power required to operate and the efficiency of this type of vacuum producer are, therefore, plotted with abscissae representing the air moved in cubic feet per minute. The vacuum produced and the power required to operate are plotted as ordinates. The curves for the Spencer machine are shown in Fig. 91. This curve is taken from a four-sweeper machine and the vertical lines numbered

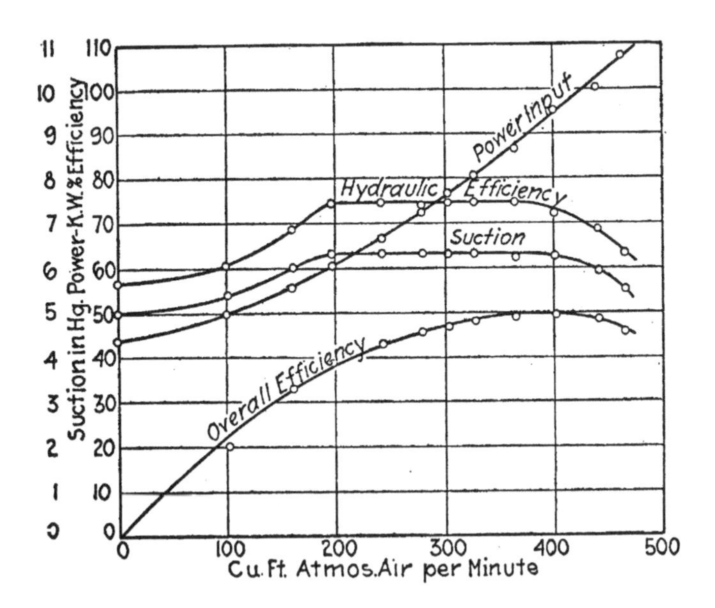

FIG. 93. POWER CONSUMPTION. VACUUM AND EFFICIENCY OF FIRST TYPES OF INVINCIBLE MACHINE.

1 to 4 represent the conditions when that number of sweepers are in operation; that is, bare floor renovators, with 50 ft. of hose or 80 cu. ft. of free air per minute. The maximum efficiency is reached at full load and is approximately 42%. The vacuum at this efficiency is 5½ in. mercury, a drop of ¾-in. from the maximum which was obtained at one-fourth load.

These machines have rather large clearances and a preliminary separator is all that is required. They operate at a speed of about 3,600 R. P. M. and the peripheral speed of the fans varies from 15,000 to 22,000 ft. per minute. This produces some noise and considerable vibration and care must be exercised in mounting the machine. In order to insure quiet running the usual method is to place the machine on a felt pad of considerable thickness.

The machines made by the Electric Renovator Manufacturing Company are horizontal and have much smaller clearances than the Spencer machines. They operate at approximately the same rotary and peripheral speed and are, therefore, as

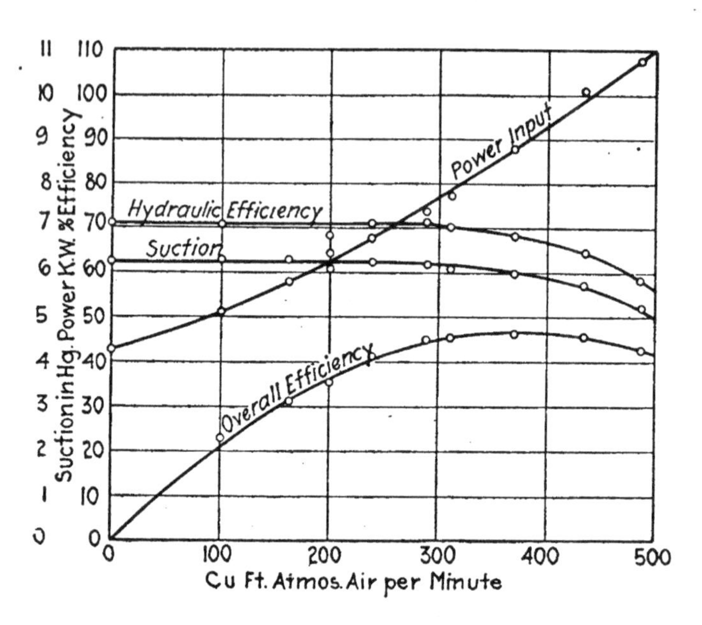

FIG. 94. POWER CONSUMPTION, VACUUM AND EFFICIENCY OF INVINCIBLE MACHINE AFTER VALVE WAS FITTED TO DISCHARGE.

noisy. However, the center of gravity of these machines is lower and the vibration is not so great. The Spencer Company is now making a horizontal machine which it furnishes only when required, the claim for their vertical machine being that the weight of the moving parts counteracts the thrust of the atmospheric pressure against the fans and relieves the work of the thrust bearings, at the expense of greater vibration. With ball bearing thrusts, the author does not consider this to be of great importance.

A view of the interior arrangement of the Invincible machine, as manufactured by the Electric Renovator Manufacturing Company, is shown in Fig. 92.

These machines, when first made, were without valves and the power consumption, vacuum and efficiency are shown in Fig. 93. It will be noted that the vacuum produced, when the machine is operated at or below one-half load, is considerably lower than is obtained at greater loads. This characteristic

FIG. 95. FOUR-SWEEPER INVINCIBLE PLANT INSTALLED IN THE UNITED STATES POST-OFFICE AT LOS ANGELES, CAL.

produces a disagreeable noise when the machine is not handling any air, evidently due to air rushing back through the outlet when the vacuum tends to build up to the maximum which occurs at intervals of about one-half second.

In order to overcome this trouble a valve has been fitted to the discharge, as indicated at 4, Fig. 92. With this valve in place the power consumption, efficiency and vacuum are as shown in Fig. 94. It will be noted that the vacuum is as high at no load as at any load up to full load and is practically constant. The efficiency at light loads is the same as before but it is slightly lower at full load, being 50% without the valve and 47% with the valve. This is due to the power being expended in opening the valve for large quantities of air and to friction in the valve passage.

A four-sweeper plant of this manufacture is shown in Fig. 95. This plant is installed in the United States Post Office at Los Angeles, Cal. The separate centrifugal separator, shown at the left of the cut, is not used in the regular equipment and was added in this case to fullfil the specification requirements.

A centrifugal pump with a single impeller is manufactured by The United Electric Company and is known as the Tuec system. A phantom view of the pump and separator is shown in Fig. 96. It will be noted that the shaft is vertical. However, the vacuum is under the impeller in this case, and the thrust due to the atmospheric pressure is down instead of up, as in the case of the Spencer machines. This throws the weight of the parts, plus the thrust due to atmospheric pressure, on the thrust bearing. These machines do not produce a vacuum greater than 3-in. mercury, and the additional thrust is not as great as in the case of the machines producing higher vacuum, the impeller being 24 in. in diameter, its area 450 sq. in. and the thrust, with a vacuum of 3-in. mercury, 675 lbs., which is worth considering. This downward thrust is partially counterbalanced by mounting the armature of the electric motor used to operate the fan, slightly below the magnetic center, thereby causing an upward magnetic pull. These machines are intended to be used with large hose and pipe lines to reduce the friction to a very low point. When operating carpet renovators the vacuum at the renovator rises to 1¾-in. mercury and the type of renovator used by them passes approximately 50 cu. ft. of air, while the bare floor renovators pass approximately 95 cu. ft. They are extensively used where bare floor work is required, their first cost being low.

The results of tests of two of these machines of four-sweeper capacity, driven by alternating and direct-current motors, respectively, are shown in Fig. 96a. These curves indicate a considerably higher efficiency with the alternating than with the direct-current motor. This is due to the low efficiency of the special high-speed direct-current motors used with all centri-

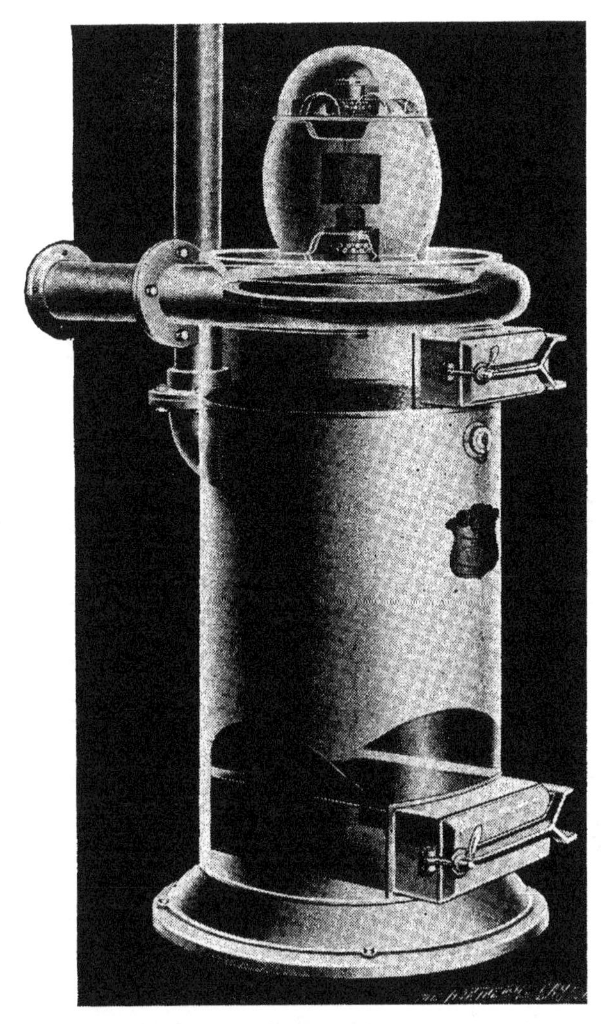

FIG. 96. CENTRIFUGAL PUMP WITH SINGLE IMPELLER, MANUFACTURED BY THE UNITED ELECTRIC COMPANY.

fugal fan-type exhausters. The alternating-current motors are not so affected, in fact, the speed at which these fans are operated is fixed by the requirements of the alternating-current motors.

The efficiency of the other types of centrifugal exhausters

(Figs. 91, 93 and 94) is in every case accomplished with direct-current motors. This machine has an efficiency about the same as the Spencer machine. It will noted that the vacuum produced does not fall off as the load increases, as in the case of the multi-stage fans. This characteristic is probably due to the fact that there is no wire drawing in the diversion vanes, as in the case of the multi-stage exhauster.

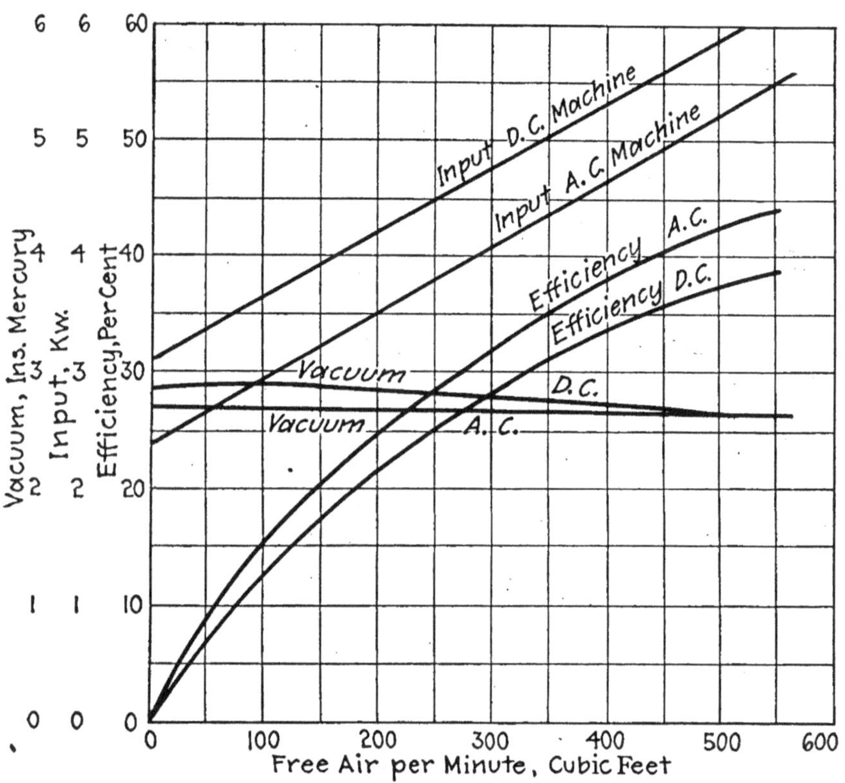

FIG. 96a. TEST OF CENTRIFUGAL PUMP WITH SINGLE IMPELLER

Steam Aspirators.— The steam aspirator as a vacuum producer in connection with vacuum cleaning systems was first used by the American Air Cleaning Company, and has been used to a limited extent by the Sanitary Devices Manufacturing Company. The type of apparatus used by the American Air Cleaning Company is illustrated in Fig. 97. A single partial separator is used with this system and the lighter dust is allowed to pass through the aspirator, where it is mixed with the steam and sterilized. The aspirator is in the form of an ejector, with a specially designed nozzle, and is always fitted with an automatic device for cutting off the steam when the vacuum in the separator reaches the degree desired.

The steam consumption required to exhaust 1 cu. ft. of free air at various vacua, as determined by actual test of four different nozzles, is shown in Fig. 98, the steam being the actual weight of dry and saturated steam at the gauge pressures noted.

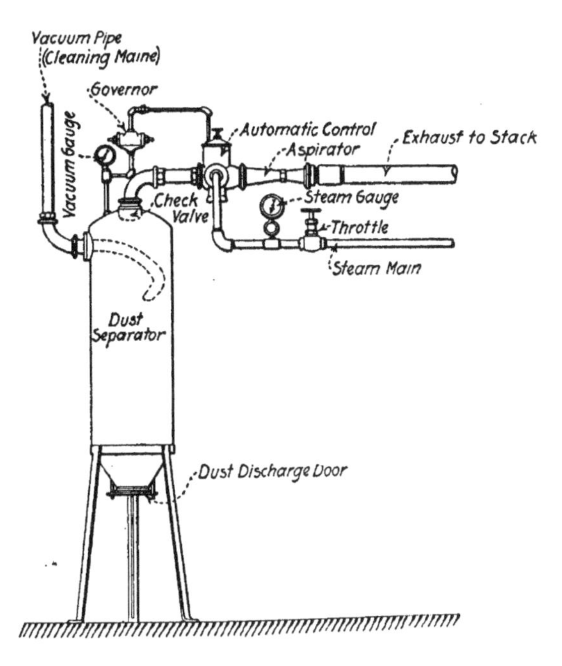

FIG. 97. STEAM ASPIRATOR USED BY THE AMERICAN AIR CLEANING COMPANY.

The American Air Cleaning Company used to guarantee a steam consumption of 250 lbs. per hour from and at 212° F., assuming that the feed water temperature was 32° F., the vacuum to be maintained at 9 in. mercury at the aspirator.

Taking the results of the test of the three-sweeper nozzle as an average, 0.066 lbs. of steam will be required to exhaust 1 cu. ft. of free air at 9 in. vacuum. The total heat in 1 pound of dry steam at 110 lbs. gauge is 1187 B. T. U. and at

212° F. the latent heat is 970 B. T. U. The factor of evaporation, therefore, is 1.235, and the weight of steam at 110 lbs. allowed by the guarantee is 202 lbs. This amount of steam will exhaust 3,060 cu. ft. per hour, or 51 cu. ft. per minute, which is more than sufficient to operate a carpet renovator, and is a little less than will pass through a bare floor brush attached to the end of 50 ft. of 1 in. diameter hose, if the hose is attached directly to the aspirator. With a line of pipe between

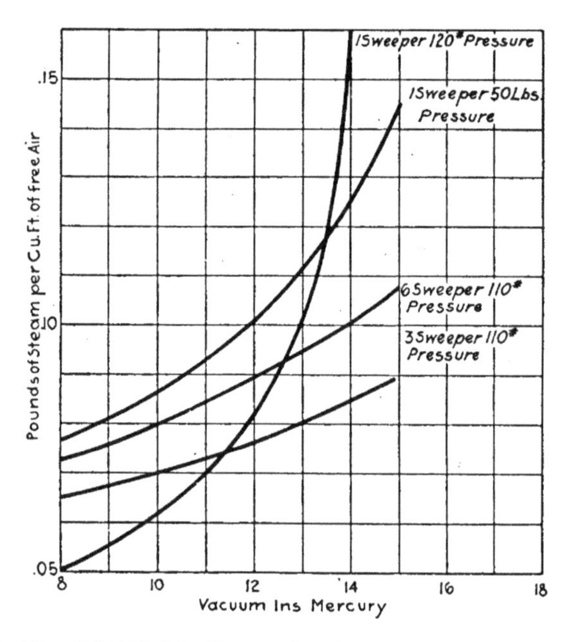

FIG. 98. STEAM CONSUMPTION OF STEAM ASPIRATOR.

the hose cock and the aspirator, the air quantity will be somewhat less, and this guarantee will undoubtedly be fulfilled in every case.

The advisability of using an aspirator will depend on the conditions to be met at the building in each case. Three typical cases are cited below:

1. When there is a Generating Plant in the Building, and a Plant Using 1¼-in. Hose and 8-in. Vacuum is Desired.—A Root blower will require 27 watts for each cubic foot of air exhausted (Fig. 88), and the three-sweeper aspirator, 0.065

lbs. of steam. Then the pounds of steam required by the aspirator to do the same work as one K. W. hour at the motor of the Root blower will be

$$\frac{0.065 \times 60}{0.027} = 146.3$$

The generating plant will produce a kilowatt hour at the switchboard with not exceeding 60 lbs. of steam, and if the transmission loss is 10% there will be required by the Root blower not over 66 lbs. of steam to do the same work that takes 146 lbs. with the aspirator. This case would require that the Root blower, driven by an electric motor, be used.

2. **When there is High Pressure Steam Available, but no Generating Plant.**—Then we may use either the aspirator or a Root blower driven by a steam engine. This engine should have an economy of 60 lbs. per indicated horse power, with not over 15% friction loss, which will require 69 lbs. per brake horse power. This will be equivalent to $69 \times 0.776 = 90\frac{1}{2}$ lbs. per K. W. hour, which is still much better than 146 lbs. required by the aspirator.

3. **When Steam is Generated on the Premises with Coal Costing $3.00 per ton and all Machinery Must be Driven by Electricity Purchased for 5 Cents per K. W. Hour.**—Cost of steam to do the same work in the aspirator that 1 K. W. hour will do in a motor driving a Root blower is:

$$\frac{146 \times 300}{7 \times 2240} = 2.8 \text{ cents}$$

as against 5 cents that would have to be paid for current. In this case there would be a saving in using the aspirator, which would not require as much attention as the motor, and at loads less than full load, the steam used by the aspirator would be in direct proportion to the load, as the control would shut the steam off entirely during a portion of the time, while the motor would require some current as long as it was in operation, even if no air was being exhausted. On the other hand, the steam which is exhausted from the aspirator is not suitable for use in heating, as it is mixed with air and fine dirt, and must be thrown away, a condition that must always be considered where there is an opportunity to use exhaust steam for heating or other purposes.

CHAPTER X.

CONTROL.

When the displacement type of vacuum producer of more than one-sweeper capacity is used with a vacuum cleaning system, some means must be employed to prevent the vacuum rising above that necessary for efficient operation of the sweepers when there are less renovators in use than the capacity of the vacuum producer or when carpet renovators are in use on all outlets.

If the displacement pump be run at constant speed, every change in the quantity of air exhausted will cause a change in the vacuum produced. This will result in inefficient operation and may result in undue effort being necessary to operate the renovator and in excessive wear on the carpets.

The earlier systems were not provided with any control and the first attempt to control the vacuum was by placing a spring relief valve on the pipe line near the separator, which admitted additional air when the vacuum tended to rise. This resulted in full load being thrown on the pump at all times when the same was in use, which does not give economical operation.

The controllers that have been devised for maintaining a constant vacuum without the introduction of air into the system operate on one of three principles:

1. Closing the suction of the vacuum producer.
2. Opening the suction of the vacuum producer and holding vacuum in the system.
3. Varying the speed of the vacuum producer.

The first type of controller was introduced in the vacuum cleaning field by the Sanitary Devices Manufacturing Company, and was known as the "unloading valve." It was similar to the unloader which had been used for some time in connection with air compressors. The detail of construction is shown in Fig. 99, and consists of a balanced valve, which is connected to a weighted piston, operating in a chamber communicating with the separators by a pilot valve. The pilot valve is operated by an auxiliary piston which is weighted to overcome the lifting effort due to the vacuum desired. When the vacuum in the cylinder becomes great enough to overcome the weights attached to the auxiliary piston, it rises, allowing vacuum to reach the main piston, which is drawn up and the suction valve closed. When this valve is closed the vacuum in the pump at once starts to build up to the maximum possible for the pump to produce, and if the pump used is of the piston type the vacuum will run up to nearly 28 in., resulting in the pump's taking the least power on which it can be operated. As soon as the vacuum in the separators falls below that which will sustain the weight on the auxiliary piston the valve falls open and the pump again draws air through the system. In actual practice this valve will operate at more or less frequent intervals. The author timed the action of one of these valves connected to the suction of an eight-sweeper piston pump, and its time varied from 2/5 second to 65 seconds. The current taken by the pump when the suction was open was 100 amperes at 220 volts. When the valve was closed for but 2/5 second the current dropped to 75 amperes, there not being sufficient elapsed time for the pump to produce a perfect vacuum. When the valve was closed for

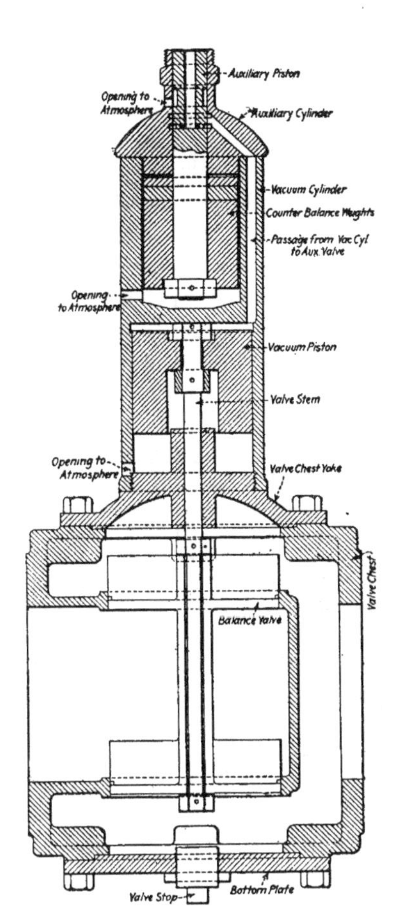

FIG. 99. FIRST TYPE OF CONTROLLER INTRODUCED BY THE SANITARY DEVICES MANUFACTURING COMPANY, KNOWN AS THE "UNLOADING VALVE."

2 1/5 seconds, the vacuum reached its maximum value and the current fell to 32 amperes.

Fig. 100 is a curve plotted from the results of this test and shows an increase in the power above that necessary to overcome the friction in the moving parts of the pump in direct proportion to the percentage of full load that the pump was serving.

This is as near an ideal condition as one could expect to obtain by any means other than stopping the pump or otherwise decreasing the friction load. However, this form of

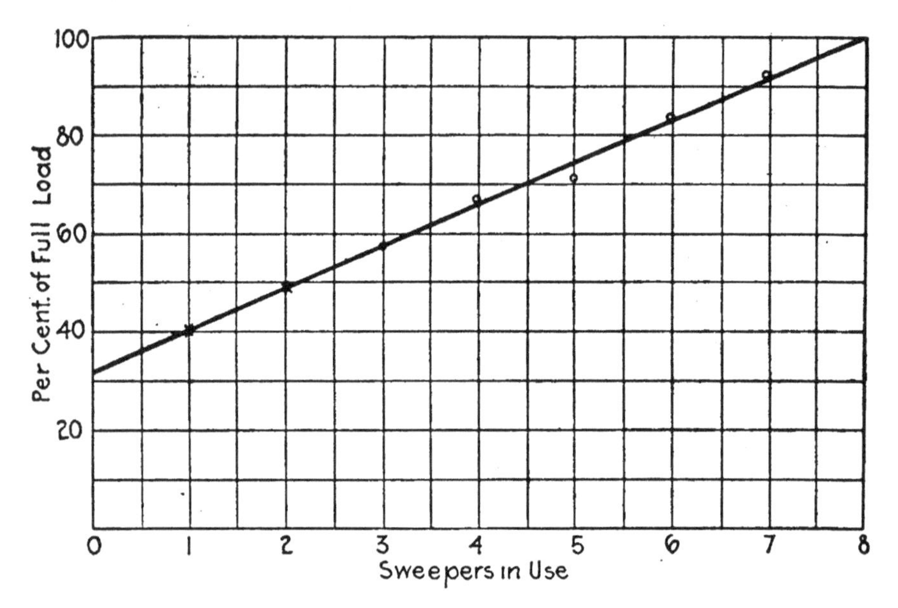

FIG. 100. TEST OF CONTROLLER CONNECTED TO SUCTION OF 8-SWEEPER PISTON PUMP.

unloader is not suitable for a pump without valves, as the power will increase with an increase in vacuum, and other means must be employed to control such a pump.

The second form of control is adapted to this type of pump. The arrangement of one of these controls is shown in Fig. 101. This consists of a single-ported valve opened by the vacuum in the cylinder, M, the action of which is controlled by a pilot or auxiliary control valve actuated by the vacuum in the separator. This auxiliary valve is fitted with two pistons, S and

O, which are held together by springs, and when so held the main cylinder is open to the atmosphere through the small ports in the piston, O. When the vacuum in the separator becomes great enough to overcome the compressive strength of the springs, T and P, the pistons, S and O, are drawn apart, closing the port in the piston, O, and opening the port in piston, S, allowing the vacuum to enter the main cylinder, M, and open the main valve. This valve permits the atmospheric pressure to enter the pump suction, the air being prevented from entering the separators by a check valve, not shown. The pump then operates without producing any vacuum, and the

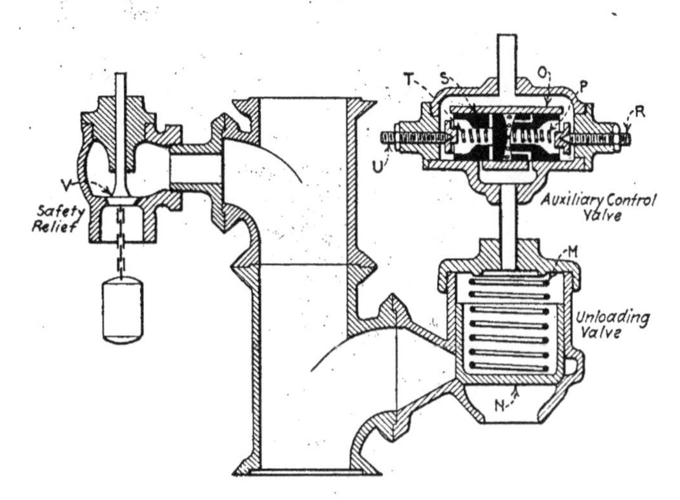

FIG. 101. TYPE OF CONTROLLER FOR USE ON PUMPS WITHOUT VALVES.

power required to operate the pump is reduced. A relief valve of the common vacuum-breaker type is shown at the left of the cut. This valve is provided to prevent overload in case the control fails to operate.

This type of control does not effect as great a reduction in the power as the first type of control described, since it requires a greater per cent. of the full load power to operate the pump at no vacuum than at perfect vacuum. No air is moved in the latter case, and the maximum volume of air is moved in the former case.

Either of these controls gives fairly economical results when the pump is serving at least a part of the sweepers at all times. However, when the system is used in a building where there may be cleaning done at any time and vacuum must be "on tap" at all times, as in a hotel, there will be many occasions when no sweepers will be in use, and the pump might then be stopped entirely, provided that it could be automatically started when needed.

FIG. 102. REGULATOR FOR MOTOR-DRIVEN VACUUM PUMP, MANUFACTURED BY THE CUTLER-HAMMER MANUFACTURING CO.

Where the steam aspirator is used, the control (Fig. 107) is attached to the steam supply valve. When the valve is closed no steam is consumed by the aspirator. This is the ideal condition where we must keep vacuum "on tap," and is a characteristic of the aspirator system which has led to its introduction in many instances.

The same economy can be obtained with a steam-driven pump by inserting a throttle valve, controlled by the vacuum in the separators, which will start and stop the engine driving the pump and vary its speed in accordance with the quantity of air required by the system.

Several appliances for varying the speed of a motor-driven vacuum pump have been placed on the market, the simplest and probably the best of these appliances being that manufactured by the Cutler Hammer Manufacturing Company, illustrated in Figs. 102 and 103.

The object of the apparatus shown in Fig. 102 is to automatically start a motor-driven vacuum pump and control the

FIG. 103. INSPIRATOR TYPE VACUUM CONTACTOR, USED TO CONTROL PILOT MOTOR OF CUTLER-HAMMER CONTROLLER.

speed of the motor so that the vacuum is maintained at the desired degree, irrespective of variation in the number of sweepers in use. This control of the degree of variation is accomplished in a more efficient manner than if the pump were to be driven at its maximum speed at all times and the pressure kept at the desired point by means of a blow-off or by-pass valve. With this system a motor is used having a control, by shunt field weakening, of approximately 3:1 in order that the control of the speed may be as efficient as possible.

Referring to Fig. 103, a small pilot motor is mounted

on brackets at the side of the panel, driving directly, through an insulating coupling, a screw shaft which carries a traveling cross-head. This cross-head is shown in the photograph at the extreme right of its travel, which corresponds to the maximum speed of the motor, the left-hand end corresponding to zero speed of the motor. In this position the motor circuit is opened by the clapper type magnetic switch. Assuming that the cross-head is in the extreme left-hand position and the knife switch is closed, the pilot motor will be started in such a direction as to move the cross-head to the right A slight movement in this direction completes a connection to the magnetic switch, which thereupon closes the motor circuit through all of the resistance, starting the pump motor.

Inasmuch as the pilot continues to move the cross-head toward the right, the speed of the pump will be gradually increased until, at a point about midway of its travel, all of the resistance in the armature circuit of the motor will have been cut out upon the upper segments and further movement then serves to weaken the field. This is accomplishd by means of the contact buttons shown just below the screw shaft.

As soon as the cross-head has weakened the field to its minimum value and thus speeded the motor up to its maximum point, a limit switch·stops the pilot motor and thus prevents further motion in that direction. As soon as the pump working this at its maximum speed has produced a vacuum in the cleaning system of, say, 12 in. of mercury, the cross-head will begin to move backward and reduce the speed to a point corresponding with the air required.

This control of the pilot motor is accomplished by means of what is termed "inspirator type vacuum contactor." This apparatus is shown more in detail in Fig. 103, and consists of a diaphgram closing one side of a chamber. The diaphragm is pressed outward by an internal spring whose tension may be adjusted by means of a hexagonal head cap screw, visible in the photograph of the complete regulator.

The diaphragm is coupled to a pivoted arm carrying insulated conical-pointed silver screws, so located that they enter holes in small silver plates mounted on opposite sides, respectively, of the upper and lower contact posts. These contact posts are hollow and communicate with the diaphragm chamber, which latter is connected by piping to the vacuum system.

Normally, the internal spring forces the diaphragm over so that the lever makes contact with the lower post. This serves to drive the pilot motor in a direction to move the cross-head to increase the speed of the pump. When the degree of vacuum for which the apparatus is adjusted is reached the lever starts to move toward the left hand, and in so doing stops the pilot motor. This maintains the pump speed at that particular value. Should the vacuum increase to a sufficient degree the lever will be drawn further over toward the left and contact will then be established with the upper post, which will cause the pilot motor to move the cross-head to the left, and thus decrease the pump motor speed.

Inasmuch as the motion of the diaphragm lever is very gradual, destructive arcing would take place at the pilot motor contacts were it not for the small openings in the silver contact plates, which, as the pointed screw leaves the hole, immediately sucks the arc inward and extinguishes it.

This method of preventing arcing is exceedingly unique and is subject to patents now pending.

It is possible to adjust the high and low limits by changing the setting of the pointed silver screws, the usual adjustments being such as to maintain the vacuum within 2 in. of mercury. The speed of the pilot motor may be adjusted by means of the small link shown in the upper left-hand corner of the panels to correspond with the capacity of the system, it being found that systems of large capacity require a slower motion than those in which the amount of piping, etc., is less for the same size of pump. In practice, the regulator will very quickly find the position corresponding to the proper speed for the number of outlets in use, and only moves a slight amount either side of this particular position.

With this regulator it is possible to employ remote control permitting the establishment of vacuum in the piping system by the turning of a pilot switch located at any point in the building. If desired, several such switches may be placed in parallel, and, under these conditions, the turning on of any switch will establish the vacuum supply which will be maintained until all of the pilot switches are turned off. By this means it is possible to have several janitors working at the same time on different floors of the building, and each will be independent of the others in his control of the vacuum; although one man may finish and turn off the switch on his floor, the pump will not be stopped if the vacuum is still required by workers on other floors.

When the total size of one installation becomes greater than 25 H. P., it is found desirable to provide two pumping units, and, in this case, the same system is applicable. The cross-head is then arranged to start first one pump and increase its speed to a maximum. If this does not supply the necessary amount of air, the cross-head continues to move, and starts the second pump, which will then be run at a necessary speed to supply the remaining amount of air.

The first pump always remains in motion at its point of highest efficiency. It is evident that this duplex arrangement is more efficient than one large pump when only a very few sweepers are in operation, since, for this condition, the very large pump would have to be run at such a slow speed that the armature resistance would be in circuit, while the single smaller pump would be running at a more efficient speed and with less proportionate motor losses.

In duplex outfits switches are provided for disconnecting either motor in case of its being necessary to clean or repair either unit. When so disconnected the other unit may be operated and maintain the same degree of vacuum within the limits of its capacity.

While this type of control is more economical in current consumption than either of the former types described, its cost is much higher, and it is seldom used unless specifically ordered.

When the centrifugal type of vacuum producers is used no control is necessary, as the inherent feature of this type of apparatus insures a practically constant vacuum at all air quantities within the capacity of the machine.

CHAPTER XI.

Scrubbing Systems.

Vacuum cleaning systems in which appliances for scrubbing are provided in addition to the usual appliances for the removal of the dust and other materials in a dry state have been introduced by a few manufacturers, none of which has come into general use.

The usual method employed is to provide an ordinary corn scrubbing brush which has a connection to the water supply of the building, with control valves in the tool handle for regulating the flow of water to the brush. Soap is applied either in the form of soap powder sprinkled on the floor, in a liquid state fed into the water supply by means of a sight-feed oil cup or soft soap in a plastic state fed into the water supply by means of a compression grease cup.

In any case, the water is run onto the floor mixed with the soap and the floor scrubbed by manipulating the corn brush, in the same manner that an ordinary corn scrubbing brush without attachments would be used.

After the dirt has been loosened from the floor, the floor may be rinsed by the application of more water. The water is then drawn up from the floor by the suction of the cleaning machine, and passes through the hose and piping system to the separator and vacuum producer. To effectively remove the water a rubber-faced tool is usually employed. In one system this rubber face is arranged to permit the corn brush to be fitted over same when scrubbing is being done, and the brush must be removed from the tool before the water can be drawn up from the floor. Other manufacturers provide a double-faced tool having the brush on the opposite side of the tool from the rubber-faced slot. By reversing the tool, scrubbing and mopping can be accomplished without the removal of the corn brush from the tool, which is more convenient for the operator.

With either of the above forms of scrubbing tools it is necessary or desirable to cut off the suction to the mopping attachment when using the corn brush, and it is also necessary to cut off the water supply to the brush when using the mopping attachment. One system, introduced several years ago, conducted the water to the brush and away from the rubber-faced mopping appliance through the same hose. This arrangement requires the use of a special three-way hose cock, which had to be manipulated frequently during the scrubbing operation, requiring the time of another person in addition to the operator, or else greatly delaying the scrubbing process by requiring the operator to constantly pass back and forth between the hose cock and the scrubbing tool. This method of supplying water also requires the use of a removable corn brush attached over the rubber mopping device.

Other forms of scrubbing appliances are provided with separate hose for the water supply and suction and with valves in the handles for controlling the suction and water supply. These valves to be efficient and quick in action are generally made self-closing, otherwise they will be short-lived, as explained in Chapter V. When springs are used to close the valves, the hand and wrist will be quickly fatigued, as stated in Chapter V.

With either of the above systems all of the scrubbing, that is, the agitation of the brush, has to be performed by the operator, as in the case of the ordinary scrubbing brush. However, the combination tool is much heavier and clumsier than the ordinary scrubbing brush, and the only advantage obtained by using this heavy and clumsy appliance is the ability to supply water without carrying it in buckets, also the removal of the dirty water after scrubbing. These appliances cannot be termed mechanical scrubbers, nor can they be classed with scrubbing machines with motor-driven brushes, such as have been recently introduced.

A real mechanical scrubbing device for use with a vacuum cleaning system was manufactured by Foster & Glidden, of Buffalo, N. Y., but was never placed on the market, although at least one is in commercial operation to-day. This machine is provided with a turbine motor operated by the air current passing through the machine. This turbine revolves a pair of scrubbing brushes turning in opposite directions. Water is fed through a separate hose, and an auxiliary air inlet is opened when the suction under the brushes is closed, in order to supply the necessary air to keep the turbine running. Mr. Foster states that he has experienced no trouble in operating this machine on from 8 in. to 12 in. of vacuum, being able to scrub and remove the dirt and water with one operation. The speed with which the work was done depended on the condition of the floor, the usual rate, as given by Mr. Foster, being from 10 to 12 yds. per minute.

Mr. Foster also states that he has not pushed the introduction of this scrubber, as he considers it so far ahead of the times as to require the education of the public in the use of the hose and ordinary vacuum cleaning tools before users would be capable of successfully operating this type of scrubber.

The author considers this condition to be lamentable if true, for until some such appliance is in commercial use scrubbing attachments to a vacuum cleaning system can never compete with the mechanical scrubbing machines now on the market, and are little if any better than the old method of scrub-brush, mop and pail, and certainly not as rapid in operation.

When the vacuum cleaning systems combine scrubbing with dry cleaning, the separator and vacuum producer must provide for the removal of water as well as air. A few manufacturers have attempted this, among which are the makers of the Rotrex system, described in Chapter IX, in which the water is passed through the pump and into the sewer under sufficient pressure to overcome the friction in the exhaust pipe through which the expelled air passes after leaving the separator. This may be sufficient to force the trap seals of the plumbing system, and, if used, the discharge connection should be made to the sewer outside the main running trap, close to the fresh air inlet. As large articles cannot be allowed to pass through the pump, a screen is necessary on the inlet side of the vacuum producer, but this may give trouble, due to the clogging with litter.

The Atwood Vacuum Cleaner Company uses a wet tank on the suction side of the vacuum producer into which the dirt and water are discharged, the air being separated and passed to the vacuum producer. When this tank becomes partly filled it is necessary to shut down the plant and empty the contents of the tank by gravity into the sewer.

This method overcomes the objections to clogged screens and forced trap seals, and all litter is discharged direct to the sewer, together with a quantity of water which is presumably sufficient to flush the litter through the sewer. The last named system is still open to two objections; first, it is not automatic, and, if neglected, the tank will fill with water and force same into the vacuum producer. With the Root type of vacuum pump this will do little harm unless a large quantity of floating litter should pass into the pump. Second, the system may be operated with dry renovators exclusively for a considerable portion of the time, in which case the contents of the separator may become of such a consistency as will not readily flush through the sewer, and stoppage of the same may occur.

Another separator of this type has recently been patented by E. B. Dunn, the originator of the Dunn Locke, in which the mud and the water are automatically discharged alternately from one of two separators, as described in Chapter VIII.

Such a separator, in which sufficient water is automatically introduced to dilute the dirt and which will automatically empty when sufficiently filled, when so constructed that it will operate continuously, is considered to be the ideal separator for use with a combined cleaning and scrubbing system. Until the mechanical scrubber and an automatically operated separator are commercially introduced the author does not consider that the use of scrubbing attachments, in connection with the vacuum cleaning system, is advisable.

CHAPTER XII.

Selection of Cleaning Plant.

We have considered in detail the various appliances which, taken together, make a complete vacuum cleaning system, but without considering their relation to one another. It now becomes necessary to choose an exact type and form of each of these appliances which will produce in combination a complete vacuum cleaning system best suited to the conditions to be met in a given installation.

In selecting a vacuum cleaning system consideration must be given to the character of the material to be removed, the kind and quality of the surfaces to be cleaned, the rate at which cleaning must be done, the extent of the cleaning system, and the cost of labor to operate the system, all of which must be considered in each step in the selection of a suitable plant.

In assembling the complete system, the author will take up the various parts thereof in the order in which they were discussed in the preceding chapters.

Renovators.—The selection of renovators is the most important step in making up a vacuum cleaning system, as the entire makeup of the system, whether good or bad, is dependent on the proper selection of these tools. The carpet renovator is generally considered first in importance, because the cleaning of carpets has nearly always been found to be the principal field of usefulness in vacuum cleaning work. This is due, perhaps, largely to the fact that from the beginning of the art of vacuum cleaning, this function of the system has been held before the eyes of the public by the manufacturers of the earlier systems. Nearly all demonstrations of cleaning systems shown to the public consist of the removal of ordinary wheat flour from a carpet. The reason for this is two-fold; first, because it is the most striking demonstration to the eye of the layman, and, second, it is the easiest to accomplish with a small air displacement and small power, which was characteristic of the apparatus made by these manufacturers.

The author was at one time of the opinion that this function of the cleaning plant was given too much prominence by builders of systems having small air displacement, and letters were sent to the officials in charge of sixteen Government buildings in which vacuum cleaning systems were installed, asking them, among other questions, whether the cleaning system was used to any extent in cleaning bare floors, of which there were large areas, both wood and marble, in the buildings in question. The plants installed were of various makes, some of which maintained 12 in. mercury at the separator and used 1-in. hose, while about an equal number of others maintained 6 in. mercury at the separator and used 1½-in. hose. The answers showed that out of the sixteen buildings the cleaner was used on bare floors in but two of the buildings. One writer, who had a plant maintaining 6-in. vacuum, provided with Type F renovators and 1½-in. hose, stated that he had tried cleaning bare floors without success, as the renovator and hose became so clogged with litter as to be inoperative. The majority stated that the cleaning system displaced brooms on carpets and rugs and several stated that the cleaning system was used to advantage in cleaning walls, cases, pigeon holes and relief work.

This indicates that for the average office and departmental building the cleaning of carpets is the most important function of the vacuum cleaner. This is also true of residence work. Schools, department stores and manufacturing buildings contain very little floor space covered with carpets, and in buildings of this chcaracter the cleaning of bare floors is of the greatest importance. In such cases the efficiency of the carpet renovator can be sacrificed to a more efficient and economical operation of bare floor renovators.

In a building where carpet cleaning is an important function of the cleaning system, the selection of the carpet renovator is most important. Of all the various types of carpet renovators discussed in Chapter III, only two need to be considered, Type A and Type F. Of these, Type A is superior in all respects except the picking up of large litter, and, unless the character of the material to be removed contains a large amount of material which can be picked up by Type F renovator that will not pass Type A, Type A renovator should always be used. Even when Type F renovators are desirable, the writer considers that the plant should still contain some Type A renovators for use in places where this unusual litter will not be encountered.

Among the bare floor renovators, described in Chapter IV, only the one having a felt face, curved to permit its running over the dirt, is worthy of serious consideration. This renovator requires an inlet or vacuum breaker to keep same from sticking to the surface cleaned, the extent of such opening being dependent on the vacuum maintained in the carpet renovators, as explained in Chapter VII.

When carpet cleaning is considered as of secondary importance to bare floor cleaning, the degree of vacuum maintained at the separators may be reduced to that which will produce a vacuum of 1 in. mercury at the bare floor renovator, allowing the vacuum maintained at the carpet renovator to be whatever the conditions of hose and pipe line will produce. Under such conditions, the area of the inrush or vacuum breaker in the bare floor renovator may be reduced considerably.

The use of brush renovators is dependent on the capacity of the air exhauster supplied, as explained in Chapter VI. If it is decided that brush renovators are necessary, then the "large volume" exhauster must be installed. The advisability of such installation is dependent on the time allowed for cleaning and the cost of the operators. In residences and small buildings where the cleaning operations can be done with one or even two domestics or laborers, very little, if any, saving in the wages of operators can be effected by increasing the rate at which the cleaning can be done. In such buildings a small-volume plant will be the most economical in first cost and operation. If such a plant is installed, the brush renovators should be omitted.

In cases where bare floor cleaning is the principal function of the cleaning system the extra quantity of air at the low vacuum necessary will not require much larger expenditure of power than that needed by the small-volume plants when maintaining sufficient vacuum for effective carpet cleaning and brush renovators should be provided with systems of this character.

Hose.—In Chapter VI it is shown that when carpet renovators are operated efficiently in combination with bare floor renovators, 1¼-in. hose will produce the best results with the lowest expenditure of power at the hose cock. In Chapter VII it is shown that with pipe lines of ordinary length 1¼-in. hose also gives the best results, with the least expenditure of power at the separator, but that in cases of exceedingly long pipe lines, 1-in. hose will be the most economical. In a system where bare floor cleaning is the principal function, the vacuum to be maintaineed at the carpet renovator is no longer considered, and for such systems the largest hose which can easily be handled will cause the least hose friction and require the lowest vacuum at the hose cock. It is, therefore, the most economical to use on such a system. The author does not recommend the use of a hose larger than 1¾-in. diameter for this type of plant.

The proper hose sizes, therefore, will be: For ordinary buildings where carpet cleaning is important, 1¼-in. diameter. For installations with unusually long lines of piping, where carpet cleaning is important, 1-in. diameter.

For all systems where carpet cleaning is of secondary importance, 1½-in. or 1¾-in. diameter.

Pipe Lines.— Pipe lines should always be as large as possible without reducing the velocity in same below 40 ft. per second, as explained in Chapter VII.

Separators.— The type of separator to be used is dependent on the type of vacuum producer adopted. Where reciprocating exhausters are used, or other type of exhauster where there is rubbing contact between the moving parts and the dust, the combination of a wet and dry separator is recommended. When rotary or centrifugal exhausters having close clearances are used, total separators with bags are recommended. When exhausters with large clearances are operated, partial separators are satisfactory.

The use of any form of apparatus contemplating the adoption

of a satisfactory scrubbing system is not considered advisable, as the author believes that separators for handling water will be improved before scrubbing becomes commercially successful. Changes in the existing separators can be made when satisfactory scrubbing appliances are placed on the market, at no greater expense than would be necessary to bring up to date any of the present systems for handling water.

Vacuum Producers.—The selection of the vacuum producer is dependent on the degree of vacuum that must be maintained to effectively operate the system selected. For the operation of a system where carpet cleaning is the principal function and 1¼-in. hose is used, the vacuum required at the producer will be from 6 in. to 9 in. mercury. Inspection of the efficiency curves of the various types of vacuum producers, given in Chapter IX, shows that the two-impeller rotary pump has the highest efficiency at this vacuum.

For the operation of systems where carpet cleaning is the most important function and 1-in. hose is found to be the most economical, 14 in. to 15 in. of vacuum at the vacuum producer is necessary, and efficiency curves, given in Chapter IX, show that the piston pump is the best suited for such service.

For the operation of a system where carpet cleaning is of secondary importance a vacuum at the producer of from 2 in. to 4 in. of mercury will be sufficient. For this work, the multi-stage or even single-stage centrifugal fan is practically as efficient as the two-impeller rotary, and will be lower in first cost and cost of maintenance. Either of the above mentioned vacuum producers are satisfactory for operating a system of this type.

Control.— Every system of more than one-sweeper capacity in which a displacement type of exhauster is used should be provided with some means of economically controlling the vacuum at the producer. On one-sweeper plants an automatic starter which will stop the motor when the vacuum reaches a point 2 in. above that required and start same when the vacuum drops to 1 in. below that required is convenient, but not necessary.

For piston pumps and all other displacement pumps fitted with eduction valves, an unloading device, which closes the suction when the necessary vacuum is exceeded, is the least expensive to install and gives very good economy when the demand on the plant is fairly continuous during the time it is in operation. Where the service is intermittent and required at nearly all hours, the Cutler Hammer control, described in Chapter X, is the most economical.

With displacement exhausters having no eduction valves, the by-pass type of control is satisfactory where the service is continuous, but is not as economical, as the unloader used with producers having eduction valves and the Cutler Hammer control is more efficient under all conditions of service. Centrifugal exhausters need no control, as vacuum control is an inherent feature of these machines.

Summing up the subject, we can divide the vacuum cleaning systems into four classes, each of which requires a different selection of appliances. They are as follows:

Class 1.—Plant for residence or small office or departmental building, to be not more than one-sweeper capacity.

Renovators: See list given for "small volume" plant, Chapter IV.

Hose: 1¼-in. diameter.

Separator: Centrifugal, dry, with bag or screen.

Vacuum Producer: Two impeller, rotary, alternate centrifugal fan. Capacity, 30 cu ft. of free air per minute, 4 in. vacuum at producer.

Control: Automatic starter, operated by vacuum.

Size of motor: ½ to 1 H. P.

Class 2.—Large office or departmental building where carpet cleaning is important and pipe lines are of reasonable length.

Renovators: See list given for "large volume" plant, Chapter IV.

Hose: 1¼-in. diameter.

Separator: Centrifugal, dry, with bag or screen.

Vacuum Producer: Two impeller, rotary. Capacity, 70 cu. ft. of free air per minute for each sweeper of plant capacity at 7 in. to 9 in. vacuum.

Size of motor: 2½ H. P. per sweeper capacity.

Control: Cutler Hammer.

Class 3.—Large building or group of buildings where carpet cleaning is important and long lines of piping are necessary.

Renovator: See list for "large volume" plants, Chapter IV.

Hose: 1-in. diameter.

Separators: One centrifugal dry and one wet.

Vacuum Producer: Piston type pump. Capacity, 45 cu. ft. of free air per minute for each sweeper of plant capacity at 14 in. vacuum.

Size of motor: 4 H. P. for each sweeper of plant capacity.

Control: Automatic unloader for continuous service. Cutler Hammer for intermittent work at all times.

Class 4.—Large or small plant where carpet cleaning is not an important function of the cleaning system.

Renovators: Same as for Class 3.

Hose: 1½ in. or 1¾ in.

Separators: One centrifugal, dry, with or without bag, according to type of exhauster adopted.

Vacuum Producer: Centrifugal fan or two-impeller rotary pump.

Capacity: 70 to 90 cu. ft. of free air per minute for each sweeper of plant capacity, with a vacuum of from 2 in. to 3 in. mercury.

Size of motor: 1 to 2 H. P. for each sweeper of plant capacity.

Control: With centrifugal fan, none; with pump, Cutler Hammer.

It is interesting to note that to produce the most efficient plant for all of the four cases named, all of the various types of vacuum cleaning systems which have come into general use have to be operated each under its most favorable conditions and the engineer should select his plant to best fulfill the conditions of the special case at hand, just as he would select his prime mover for an electric generating plant according to its size and location. There should be no more reason why any one of these systems should attempt to fulfill the requirements of every installation than there would be for a manufacturer of steam engines to attempt to use the same type of engine to

drive a generator under all conditions.
this condition will soon be realized by
vacuum cleaning systems and that they
apparatus of the type best suited to the
each case.

CHAPTER XIII.

TESTS.

Having decided on the type of vacuum cleaning system that is best suited to the conditions of the particular building in which it is to be installed, it then becomes necessary to ascertain what are the tests necessary to determine whether the installation will produce the desired results.

If the installation is one in which carpet cleaning is important and the plant is of more than one-sweeper capacity, the exhauster must be of sufficient capacity to produce a vacuum of not less than 4 in. mercury at a carpet renovator attached to any inlet on the piping system, when the plant is operating other renovators of any type attached to any of the other inlets corresponding to one less than the total sweeper capacity of the system.

When hose lengths as short as 25 ft. can be used on any or all of the outlets, it has been demonstrated in Chapter VII that an air removal of 70 cu. ft. of free air per minute for each sweeper of plant capacity is necessary, no matter what size of hose is used. It was also shown that where pipe lines are very long and it is possible to always use 100 ft. of hose, efficient cleaning can be done with less expenditure of power with an air displacement of 45 cu. ft. of free air for each sweeper of plant capacity.

Many methods have been recommended for testing a cleaning plant. Perhaps the earliest was the maintaining of 15 in. of vacuum at the vacuum producer with carpet renovators each attached to 100 ft. of hose, equal in number to the sweeper capacity of the plant in operation on carpets. Another test is to attach 100-ft. lengths of hose to inlets on the system, with the ends wide open, equal in number to the sweeper capacity of the plant, and require the pump to maintain a vacuum of 15 in. mercury.

Both of these tests were recommended for use on plants where 1-in. diameter hose was provided and the results are dependent largely on the size and length of the piping system. With an average-sized system, the first test will require an exhaustion of approximately 25 cu. ft. of free air per renovator per minute if Type A renovators are used. The second test will require an exhaustion of approximately 50 cu. ft. of free air per open hose per minute. Neither of these tests will insure a plant of sufficient capacity to do effective cleaning where 25-ft. lengths of 1-in. hose can be used or if larger bore than 1-in. hose be used.

If these tests are required with bores larger than 1-in. diameter and the vacuum is maintained the same as before, air exhaustion with 1¼-in. open hose will be approximately 70 cu. ft. of free air per open hose, and with 1½-in. hose, approximately 150 cu. ft. per open hose, while, if carpet renovators be used, the vacuum at the renovator would be from 7 to 9 in. of mercury. In either case, the vacuum required to be maintained at the separators is higher than is necessary to produce economical cleaning with either 1¼-in. or 1½-in. hose.

Tests with carpet renovators attached to 100 ft. hose lines in number equal to the capacity of the plant, and a vacuum of 4½ in. of mercury at the renovator will result in an exhaustion below that necessary to produce efficient cleaning when bare floor renovators and carpet renovators with shorter hose lines are used, as is likely to occur in actual practice.

Again, open hose tests require a variable length of hose to be used in order to obtain the same air quantity with the proper vacuum at the separator for economical operation.

If 70 cu. ft. of air is desired, as in the case of Class 2 plant (Chapter XII), the hose lengths should be:

50 ft. 1 in. diameter. 12 in. vacuum at separator.

75 ft. 1¼ in. diameter. 9 in. vacuum at separator.

125 ft. 1½ in. diameter. 6 in. vacuum at separator.

Any of these lengths would give satisfactory cleaning with one carpet renovator in use, together with sufficient bare floor renovators to equal the capacity of the plant. This is a possible condition in any plant.

Another method of testing is to measure the actual air passing through a given length of hose and require sufficient vacuum at the separator to produce this flow. This method is open to the objection that variation in the size of the hose will result in considerable variation in the vacuum at the separator and conditions of hose lengths may be such that when carpet renovators are attached to the hose, the vacuum at the renovator will vary according to the resistance offered to the passage of the air by the friction in the hose. With the small hose, the friction will be greatest, and the reduction in the quantity of air passing the renovator from that passing an open hose will result in the greatest reduction in friction loss through the hose and produce the highest vacuum at the renovator. This will cause a widely different vacuum at the renovator with different sizes of hose, each of which passes the same amount of air with the end of hose open.

What is desired in cleaning operations is a certain degree of vacuum at the carpet renovator, with the system operated under the same conditions that will obtain in practical cleaning, and with cleaaners of various types attached to hose ends equal in number to the capacity of the plant.

The most rational system of testing is one in which the actual conditions of air quantity and vacuum are measured at the hose ends. This can be obtained by actually attaching cleaning tools to the hose ends and measuring the vacuum within the renovator. However, a wide variation in vacuum will result when the renovator is moved along the carpet, and this variation will be different with different operators and different grades of carpet to such an extent as to render it impossible to actually meet any reequirements that may be specified, unless a considerable variation in vacuum is permitted.

It is also possible for an operator to become so expert in the manipulation of the renovators as to be able to meet the specification requirements with a plant which will not give satisfactory results in actual operation.

The most satisfactory method of testing that has been devised is the use of an orifice of proper size fixed to the hose end and measure the vacuum just inside of this orifice. In making such measurements care must be taken that the tube connecting to the vacuum gauge is not inserted in such a manner that the air velocity affects the reading of the vacuum gauge. The shape of orifice must also be carefully specified, as the rounding of the edges of the opening will greatly increase the quantity of air passing a given-sized orificee. The best standard is a sharp-edged orifice in a thin disk which has a coefficient of ingress of approximately 65%.

A convenient form of testing appliance based on the orifice test is the vacometer, manufactured by the Spencer Turbine Cleaner Company and shown in Fig. 104. This device consists of a spherical aluminum casting, with a 1-in. diameter hole on the equatorial circle, a vacuum gauge being attached to one polar extremity, the other being attached to the end of the hose. A ring having a slip fit is placed around the equatorial circle in which openings varying from ½-in. to ⅞-in. diameter are drilled. By turning this ring any of the orifices may be made to register with the opening in the sphere. The opening to which the vacuum gauge is attached is so located that it is not affected by the entering air current, and its readings are not affected by the velocity head.

FIG. 104. VACOMETER FOR USE IN TESTING VACUUM CLEANING SYSTEMS.

Experiments with this instrument in connection with a Pitot tube show that a ½-in. diameter orifice is equivalent to a Type A carpet renovator, a 5/8-in. orifice to a Type F renovator and a ⅞-in. orifice to a bare floor renovator.

With instruments of this type equal in number to the capacity of the plant in sweepers, attached to the ends of the cleaning hose, it is possible to obtain uniform conditions equal to the average results that will be obtained in actual practice with renovators attached to the hose, without the possibility of expert manipulation of the renovators affecting the results.

The proper orifice to be used in each vacometer during the test will vary with the character of the service for which the plant is designed, and the author recommends the following for each of the classes of plants described in Chapter XII:

Class 1. 2-in. mercury, with ½-in. orifice, maximum length of hose to be used in actual cleaning.

Class 2. One-half the inlets ½-in. orifice, 4.5 in. mercury at one orifice attached at end of longest hose desired to use in practice, the remaining ½-in. outlets on shorter hose lengths. The other half of inlets to have ⅞-in. orifices open at same time, with longest hose on one-quarter of total inlets and shortest on the balance.

Class 3. All inlets on long hose, one-half with ½-in. orifice, balance with ⅞ in.

Class 4. All inlets to have ⅞-in. orifice and 1 in. vacuum at vacometer, all hose lines maximum length.

CHAPTER XIV.

SPECIFICATIONS.

Before the engineer begins to prepare his specifications for a proposed vacuum cleaning system, he will naturally consider carefully the conditions to be met in the particular installation contemplated. Having considered these conditions, he can readily determine the type of system that will operate most efficiently and economically under such conditions. It is, therefore, natural to assume that the best interests of his clients can be obtained by confining his specifications to apparatus of the type giving the most efficient results for the special conditions to be met. However, it is also necessary to study the apparatus on the market to determine if there is a sufficient number of manufacturers producing the particular type of apparatus specified to insure healthy competition and reasonable bids.

It becomes necessary, therefore, to examine the various systems offered by the manufacturers in order to determine what competition can be obtained.

Apparatus for Class 1 or Class 2, if confined to the positive displacement rotary exhausters of the two-impeller type, can be obtained from at least seven manufacturers. If the centrifugal fan is included, at least three other manufacturers can be considered and in either case a healthy competition be had.

If apparatus of Class 3 is desired, it can be obtained from at least three manufacturers. A few years ago more manufacturers of systems of this type were in the market. Some of these have dropped out, owing to the comparatively limited field for this apparatus. However, there are still enough manufacturers in the field to insure competition.

Apparatus of Class 4 has been especially manufactured by one company. However, any of the manufacturers of centrifugal fan type of apparatus can easily meet the specification requirements for apparatus of the character.

It is, therefore, evident that the specification of apparatus of the type best suited to any particular installation will not result in lack of competition, and such a procedure would apparently be justified.

There are installations, such as those for public buildings, where it may be advisable from an administrative standpoint to allow the widest competition possible. In such cases the engineer can secure the best results for his clients by so drawing his specifications as to include all types of apparatus, fixing carefully the test requirements to be met and requiring each bidder to state in his proposal the amount of power required to operate his apparatus under full load, three-quarter load and half-load conditions, and to base the award of the contract on an evaluation basis.

To determine what the basis of this evalution shall be it is first necessary to ascertain the length of time the plant will be operated at each of the loads specified and find the annual cost of a unit of power to operate the plant. Assuming the plant has a life of ten years, we can charge 10% depreciation, add to this 5% for interest on the investment and 1% for insurance. We can capitalize the saving in power at 16% and use this amount as a basis for evaluation.

As an example, assume one bidder guarantees a power consumption of 1 K. W. less at full load, 1.25 K. W. less at three-quarters load and 0.75 K. W. more at half load than a lower bidder. Assume the plant will operate 500 hrs. per year at full load, 200 hrs. at three-quarters load and 300 hrs. at half load. The total kilowatt hours saved by the more economical plant will be:

Full load $500 \times 1 =$ 500
Three-quarter load $200 \times 1.25 =$ 250

Total saving 750
One-half load, $300 \times 0.75 =$ 225

Net saving (K. W. Hr.) 525

If power costs 5 cents per K. W. Hr. the yearly saving will be $26.25, which, capitalized at 16%, will equal $164.00. This is the amount which the owner would be justified in paying for the more economical plant above the price asked for the cheaper, but less economical, system.

In order to guard against any bidder guaranteeing a lower power consumption than he can actually show on test, it is necessary to impose a penalty for failure to meet the guarantee which is in excess of the increase in price shown to be justified by the evaluation.

The author recommends that this penalty be made not less than 150% of the increase in price shown by the evalution.

Actually, the owner will not lose by the less efficient plant any more than the amount shown by the evaluation if he junks the plant at the end of ten years. However, it is more than likely that he will either use it for a longer time or will be able to realize something for the plant when it is displaced. The increased penalty, therefore, is justified, and it is absolutely necessary to make this penalty greater than the increased value to prevent the bidder guaranteeing a power consumption lower than he can show on test.

The following pages contain sample specifications for apparatus of each of the four classes of systems described in Chapter XII and a specification permitting the widest competition, with evaluation and penalty clauses.

CLASS I.

PLANT FOR RESIDENCE OR SMALL OFFICE BUILDING OF ONE-SWEEPER CAPACITY.

1. *General Description.* The work included in this specification shall be the installation of a complete vacuum cleaning system for the removal of dust and dirt from rugs, carpets, floors, stairs, furniture, shelves, walls and other fixtures and furnishings throughout the building, and for conveying said dust and dirt to suitable receptacles located where shown, together with all of the necessary cleaning tools, hose, piping,

separators, exhauster, motor, etc., as hereafter more fully specified.

2. *Exhauster.* The exhauster in all of its details shall be made of the best materials suitable for the purpose and shall be of approved design and construction, and may be either of the positive displacement (rotary) or of the multi-stage fan type.

3. *Rotary Exhauster.* The rotary displacement exhauster shall be either of the two-impeller type or of type having single impeller without sliding vanes revolving without friction contact with case and with oscillating follower.

4. Exhausters fitted with sliding blade or blades will not be acceptable.

5. All parts of the exhauster shall be rigid enough to retain their shape when the machine is working under maximum-load conditions.

6. The impellers must be machined all over and must be of such shape and size that they will revolve freely and not touch each other, the follower, or the casing (cylinder) in which they are placed, but the clearance must be of the least possible amount consistent with successful operation.

7. The shafts must be of steel with the journals ground to size.

8. The journal boxes must be long and rigidly supported by the headplates and placed far enough from the headplates to allow the placing of proper stuffing boxes on the shafts.

9. The shafts of two impeller exhausters must be connected by wide-faced steel gears, cut from the solid and securely fastened to the shafts. Follower shaft on single impeller exhauster to be connected to impeller shaft by crank and connecting rod. The gears shall run in suitable oil-tight gear boxes that shall be fitted with adequate and suitable means for lubrication.

10. *Centrifugal Fan Type.* The centrifugal fan exhauster to be so proportioned and constructed as to handle the volume of air required at the specified vacuum with the least possible loss. The housing shall be of cast iron or aluminum, made in sections. The housing must be air-tight.

11. The fan wheels to be constructed of steel or other metal of high tensile strength, properly reinforced, and, if cast, must

include hub and arms complete in one piece. If the fan wheels are built up, they must be strongly riveted to cast-iron, steel or brass hubs or spiders.

12. The fan wheels are to be secured to shaft with a feather and set screws, or with left-hand screw.

13. The shaft of fan exhauster may be vertical and the wheels so mounted that their weight will equalize or partly equalize the end thrust, or the end thrust may be balanced by the magnetic pull of the armature. Shaft may be horizontal and end thrust taken care of with ball-bearing thrust rings.

14. The journal boxes for all of the above named types of exhausters shall be of the design best adapted for the purpose and must be fitted with first-class approved continuous lubricating devices, either sight feed, ring oiler, or any other kind best suited for the work or design of apparatus used.

15. *Cooling.* The rotary type of exhauster must be provided with the necessary water connections to properly seal and cool the pump. Fan type of exhauster must be designed to operate continuously without a rise of temperature over 100° F. above room temperature.

16. *Speed.* Rotary exhausters shall not exceed a peripheral speed of 1,100 ft. per minute at tips of impellers.

17. Centrifugal fans shall not exceed peripheral velocity of 22,000 ft. per minute when running under specified full-load conditions.

18. *Mounting.* The exhauster, motor and separators shall be mounted as a unit on suitable cast-iron base plate, either mounted on legs or resting on the basement floor.

19. *Drive.* The exhauster shall be driven by an electric motor, which may be direct connected to the exhauster shaft or be operated with an oak-tanned leather belt, or by cut gearing. Belt and gearing are to be of ample size and strength for their work and must run without undue noise or wear. Means shall be provided to take up the slack of the belt. Furnish and place a suitable metal guard over belt and pulley wheels that shall prevent oil being splashed outside of the base plate and prevent clothing being caught.

20. If the exhauster is operated through cut gearing, the gear-

ing must be inclosed in an oil and dust proof case, which shall be fitted with means for copious and continuous lubrication of same.

21. *Finish.* The air exhauster and motor and the base plate shall be finished in a first-class manner, filled, rubbed down and painted at least one coat at the shop, and after installation shall receive two more coats, finishing tint to be as directed.

22. *Electric Motor.* Motor to be of such size that when operating under test conditions it will not be under less than three-fourths nor more than full-load condition. It is to be of standard make, approved by the architect.

23. Motor to be wound for .... volts direct current.

24. Armature to be of toothed-core construction, with windings thoroughly insulated, and securely fastened in place, and must be balanced both mechanically and electrically.

25. Commutator segments must be of drop-forged or hard-drawn copper of highest conductivity, well insulated with mica of even thickness and proper hardness to insure uniform wear, and shall run free from sparking or flashing at the brushes under all conditions of speed. It must be free from all defects and have ample bearing surfaces and radial depth as provision for wear.

26. Brushes to be of carbon, mounted on a common rocker arm for motor, and to have cross-sectional area of not less than 1 square inch for each 35 amperes of current.

27. Brush holders to be of a design to prevent chattering, with individual adjustment in tension for each brush.

28. Bearings to be of an approved self-oiling or ring type.

29. There must be an insulation resistance between motor frame and field coils, armature windings and brush holders of not less than 1 megohm.

30. Motor must be capable of standing a breakdown test of 1,500 volts alternating current. Either or both of the foregoing tests to be applied at the discretion of the architect's agent at the time of shop tests.

31. The maximum rise in temperature of the motor at a continuous run (after installation at building) at full speed and full-rated load for a period of eight hours must not exceed 50°

C. in windings and 55° C. on commutator above the surrounding atmosphere.

32. Motor to be finished in a first-class manner, filled and rubbed down and painted two coats at the shop, and after installation to have two more coats; finishing tints to be as directed by the superintendent of the building.

33. *Tablet.* Furnish and mount where directed a polished slate tablet not less than ¾ in. thick, having mounted thereon one 30-ampere, 250-volt, double-pole knife switch, with enclosed indicating fuses, and, if displacement exhauster is furnished, one automatic self-starter having butt contacts, cutting out starting resistance in not less than two steps, starter to be controlled by the vacuum in separator, and shall stop motor when vacuum rises 2 in. above that required to meet test requirements and start motor when vacuum falls to that required for working.

34. *Electrical Connections.* This contractor shall run feeders from vacuum cleaner panel in switchboard where shown to the motor panel and make all electrical connections between panel and motor, etc.

35. All wires are to be run in standard steel conduit, except those that are so short as to be self-supporting, and these are to be cord wrapped or otherwise protected. No wire smaller than No. 8 to be used.

36. All material and workmanship to be strictly first class. Electrical work must show an insulation resistance of at least 1 megohm, and to be in strict accordance with the latest edition of the "National Electrical Code."

37. *Dust Separator.* There shall be one dry separator located where shown on plans, having a volume not less than 3 cu. ft.

38. The interior arrangement of the separator shall be such that no part of same will receive the direct impact of the dust. Cloth bags or metal screens if used in this separator shall be so placed that nothing but the very lightest of the dust can lodge thereon, and that same may be easily cleaned without dismantling the separator. It must be so constructed that it shall intercept not less than 95% of the dust entering same.

38a. Separator tank shall be constructed with steel shells, with either cast iron or steel heads, and be fitted with suitable bases or floor stands for support and proper openings for cleaning same. Separator shall be fittted with iron-column mercury gauge reading 50% in excess of operating vacuum.

39. *Pipe Lines.* All pipe lines shall be of the sizes and run as indicated on drawings.

40. *Pipes.* All pipe conveying air is to be standard black wrought-iron or mild-steel screw-jointed pipe, and is to be smooth inside, free from dents, kinks, fins, or burs. Ends of pipe to be reamed to the full inside diameter and beveled. Bent pipe to be used in mains where necessary and where shown on plans.

41. Care must be taken in erecting pipe to maintain as nearly as possible a smooth, uniform bore through all pipe and fittings.

42. *Fittings.* All fittings to be tough gray cast iron, free from blowholes or other defects; smooth castings in all cases.

43. All fittings on vacuum lines must have inside diameter through body of same size as pipe bore, and all fins, burs, or rough places must be removed.

44. Fittings on vacuum lines are to be black or may be galvanized.

45. Where space permits, all tees and elbows must have a radius at center line of not less than 3 in.

46. Horizontal overhead pipes to be supported with substantial pipe hangers spaced not more than 10 ft. apart.

47. The hangers must have an approved form of adjustment and the instructions of the superintendent in regard to securing hangers to floor construction, etc., above must be carefully followed.

48. Where exposed pipes pass through walls or floors of finished rooms they must be fitted with cast-iron nickel-plated plates.

49. *Clean-Out Plugs.* Brass screw-jointed clean-out plugs are to be provided in lines at all turns where indicated on the drawing. The clean-out plugs to be 2 in. diameter, except in the 1½-in. lines, where clean-outs are to be same diameter as the lines.

50. *Exhaust Connection.* Exhaust pipe from the exhauster

is to be run up to the basement ceiling and along same into the smoke breeching beyond damper as directed.

51. *Sweeper Inlets.* The following number of inlets are to be provided: Subbasement , basement , first story , second story , attic .

52. The sweeper inlets are to be fitted with hinged covers or caps with rubber gaskets arranged to be self-closing when hose is removed, and those in corridors and lobby arranged to be opened with a key.

53. Inlets coming through finished walls or partitions are to be flush pattern.

54. Inlets on risers run exposed against walls are to be set close up against bead of fittings.

55. If contractor desires to use other form of connection than above described which is equally satisfactory, same must be submitted to the Architect for approval after award of the contract.

56. In this specification the word "renovator" is used to mean that portion of the tool which is in contact with the surfaces cleaned; the word "stem," that portion connecting the renovator and hose; the word "cleaner" is used to mean a complete cleaning tool.

57. The following cleaning tools are to be furnished:

One carpet renovator, with cleaning slot ¼ in. by 12 in. long.

One bare floor renovator, 12 in. long, with curved felt-covered face.

One wall renovator, 12 in. long, with cotton flannel curved face.

One upholstery renovator, with slot ¼ in. by 4 in.

One corner cleaner.

One radiator cleaner.

One hat brush.

One long curved stem about 5 ft. long.

One extension tube about 5 ft. long.

58. The renovators for carpets, bare floors and walls to be arranged with adjustable swivel joint, so that same can be operated at an angle with stem from 45° for regular use to an angle of about 80° for use under or back of furniture and other similar places. This movable joint to be so arranged that lips

of cleaning tool will always remain in contact with surface cleaned, and constructed so that fitted surfaces are not exposed to dust, and the air currents when deflected to impinge only upon surfaces which are of heavy metal and where such wear as occurs will not affect the operation and handling of the tool.

59. All renovators and stems are to be as light as is consistent with strength and ability to withstand cutting action of dust.

60. The lips of carpet renovators and upholstery cleaner to be of such proportions and form as will prevent injury to the fabric, and such widths as will reduce to a minimum the sticking of renovator face to the material being cleaned.

61. Stems to be not less than 1 in. outside diameter. Air passages in swivels to be same diameter as inside of stem. Stem for use with floor renovators shall be curved near upper end to form handle and provided with swivel to permit hose hanging vertical.

62. Stems to be drawn-steel or brass tubing, not less than No. 21 United States standard gauge thick if steel and not less than No. 16 Brown & Sharpe gauge thick if brass.

63. Carpet renovators to be made preferably of pressed steel, as light as possible, or may be made of cast iron, brass or aluminum with iron wearing face.

64. Bare floor renovators shall have renewable elastic wearing face curved in direction of motion when cleaning.

65. All renovators and brushes must be provided with proper rubber or other approved buffers to prevent marring the woodwork.

66. Upholstery cleaners are to have inlet slots or openings of such size and form as to absolutely prevent drawing in loose covering of furniture.

67. Upholstery and corner cleaners are not to be arranged for use with stems, but are to have their own handles permanently attached and be provided with hose couplings.

68. All metal parts of renovators and stems are to be finished, and all except aluminum parts nickel plated.

69. *Hose.* Furnish 75 ft. cleaning hose in three 25-ft. lengths.

70. The hose to be 1¼ in. inside diameter best quality rubber

hose, reinforced in best manner to absolutely prevent collapse at highest vacuum obtainable with the exhauster furnished and to prevent collapse if stepped on. Weight of hose to be not over 12 oz. per linear foot.

71. Couplings for hose to be either slip, bayonet-lock or all-rubber type, with smooth bore of practically same diameter as inside of hose. The couplings to have least possible projection outside of hose dimensions and be well rounded, so as not to injure floors, doors, furniture, etc.

72. Bayonet joints may have packing washer, and slip joints to have permanent steel pieces on ends of hose and brass slip coupler. All ends of hose at couplings to have outside ferrules securely fastened in place, or pure gum ends glued to coupling. Simple conical slip joints slipped into ends of hose without ferrules will not be acceptable. All joints must fit together so that they will not be readily pulled apart.

73. *Tests.* All piping to be tested with air pressure equal to 5 in. mercury before being concealed in walls and other spaces. Mercury must not fall more than ¼ in. in one-half hour.

74. On completion of plant the pump will be operated with all outlets closed and, under these conditions, there must be an interval of not less than 10 min. between the stopping and starting of the motor by the automatic control, if pump system is used. And if fan system be used, the power required to operate the exhauster must not be more than 65% of that required in capacity test.

75. To test the capacity of the separator, a mixture containing 6 lbs. of sand, passed through a 50-mesh screen, 3 lbs. of common wheat flour and 16 lbs. of Portland cement shall be spread over 50 sq. ft. of floor and picked up with a renovator attached to the end of 50 ft. of 1¼-in. hose. The machine shall be stopped and the material removed from the separator spread on floor and picked up. This procedure shall be repeated until the material has been handled four times. If the separator contains a bag, the same must not be disturbed until after completion of the capacity test, which will be made with the material in place in separator, after being picked up the fourth time.

After completion of capacity test, the contents of separator shall be weighed and if same be a partial separator it must contain 95% of the material picked up. If a displacement machine is used as a vacuum producer, the separator must prevent the passage of any dust through separator, which will be determined by holding a dampened cloth over pump outlet during test of apparatus. Said cloth must not show any dust lodged thereon at end of test.

76. To test the capacity of the plant a standard vacometer, attached to the end of 75 ft. of cleaning hose shall show a vacuum of 2 in. mercury with ½-in. diameter orifice open.

77. *Test of Cleaning Tools.* The plant shall be operated by the Contractor in the presence of the Architect's representative, and a test made of each kind of cleaning tool furnished. The tool shall be attached to a 50-ft. length of hose attached to an outlet selected by the Architect's representative, and under normal working conditions each tool must satisfactorily perform the work for which it was designed. Dust and surfaces to be cleaned shall be furnished by the contractor.

78. *Painting.* After the completion of the specified tests, all exposed iron work except galvanized iron or tinned work in connection with this apparatus, not specified to be otherwise finished, shall be primed with paint suitable for surfaces covered, and then given two additional coats. Machinery shall be painted as already specified, and all other work shall be given finishing tints as selected or approved by the architect. Black iron pipe, etc., shall be given two coats lead and oil of tint directed.

Modifications of Specifications when Alternating Current is Available.—When alternating current is available, instead of direct, modify specifications as follows:

23. Motor to be wound for .... volts, ...cycle, ...phase alternating current.

24. Motor to have rotor of the squirrel cage type.

Omit 25, 26 and 27.

28. To remain as for direct current.

29. There must be an insulation between the starter or primary windings and the frame of not less than one megohm.

30, 31, and 32. Same as for direct current.

33. *Tablet.* Furnish and mount where directed a polished slate tablet having mounted thereon a 30 ampere, 250 volt, ... pole knife switch with enclosed indicating fuses and, if displacement type exhauster is furnished, an automatic starter of the "across the line" type, operated by vacuum in the separator which will stop motor when the vacuum in the separator rises 2 in. above that required to meet test conditions, and start exhauster when vacuum reaches working range.

CLASS 2

Plant for Large Office Building Having Pipe Lines of Moderate Length.

1. Same as for Class 1.

2. Omit centrifugal fan.

3 to 9. Same as for Class 1.

Omit 10 to 13.

14 and 16. Same as for Class 1.

15. Omit centrifugal fan.

Omit 17 and 18.

18a. *Base Plate, Foundation, etc.* Provide suitable base plate to rigidly support the exhauster and its motor as a unit, which shall be large enough to catch all drip of water or oil. Provide a raised margin and pads for feet of exhauster frame, motor, and anchor bolts, high enough to prevent any drip from getting into the foundation or on the floor.

18b. Provide suitable foundation of brick or concrete, to which the base plate shall be firmly anchored. The foundation shall be built on top of the cement floor of the basement, which shall be picked to afford proper bond for the foundation.

18c. Construct the foundation of such a height as to bring the working parts of the machine at the most convenient level for operating purposes. Exposed parts of the foundation to be faced with best grade white enameled brick. If the base plate does not cover the foundation, the exposed top surface is to be finished with enameled brick, using bull-nose brick on all edges and corners.

19 to 23. Same as for Class 1.

23a. The guaranteed efficiency of motor shall not be less than 78% at half load and not less than 84% at full load.

24 to 32. Same as for Class 1.

32a. Motor shall be subject to shop test to determine efficiency, heating, insulation, etc. Manufacturer's certified test sheets of motor giving all readings taken during shop tests, together with calculated results, must be submitted to the Architect for approval before motor is shipped from factory.

33. *Rheostat.* Furnish and install where shown, upon a slate panel hereinafter specified, a starting rheostat of proper size and approved made, designed for the particular duty it has to perform. It must have an automatic no-voltage and overload release. All resistance for rheostat is to be placed on the back of the tablet. Contacts must project through board to front side. All moving parts must be on front of board.

33a. *Tablet.* Furnish and place where shown, a slate tablet not less than ¾ in. thick, supported by a substantial angle iron frame, so placed that there will be a space of not less than 4 in. between the wall and back of resistance. Mount on this tablet one double-pole, 250-volt knife switch, with two 250-volt inclosed fuses and one starting rheostat, as specified hereinbefore. The connections shall be on the back of the tablet. The space between the column and the tablet shall be inclosed with a removable diamond-mesh grill of No. 10 iron wire in channel frame.

34, 35, 36. Same as for Class 1.

36a. *Automatic Control.* Suitable means shall be provided in connection with the rotary exhausters that will maintain the vacuum in the separators within the limit of the machine at point found to be most desirable, irrespective of the number of sweepers in operation.

36b. Controller shall consist of a suitable means provided in the exhauster, or as an attachment thereto, which will automatically throw the exhauster out of action by admitting atmospheric pressure to the exhauster only, but not to the system whenever the vacuum in the separators rises above the point considered desirable, and throw the exhauster into action when the vacuum falls below the established lower limit.

36c. *Vacuum Breaker.* In addition to the controlling devices above specified there shall be placed in the suction pipe to the exhauster an approved positive-acting vacuum breaker having opening equivalent to the area of 1-in. diameter pipe and set to open at 10 inches vacuum.

(If plant is to be run for long periods without much load, as in a hotel, omit 36a, b, c, and substitute):

36d. *Automatic Control.* An approved type of controller for maintaining practically a constant vacuum by varying the speed of the motor driving exhauster arranged to permit the operation of the motor continuously at any speed between full speed and stop, so long as there be no change in vacuum and which will increase speed whenever vacuum falls and reduce speed whenever vacuum rises, must be provided.

37. *Dust Separator.* There shall be one dry separator located where shown on plans, having a volume of not less than 3 cu. ft. for each sweeper of plant capacity.

38. The interior arrangement of the separator shall be such that no part of same will receive the direct impact of the dust. Cloth bags or metal screens, if used in this separator, shall be so placed that nothing but the very lightest of the dust can lodge thereon, and that same may be easily cleaned without dismantling the separator. It must be so constructed that it will intercept all of the dust entering same.

38a to 56. Same as for Class 1.

56a. *Tool Cases.* Furnish approved hardwood cabinet-finished cases for cleaning tools. Each case to be made as light as possible and of convenient form for carrying by hand and provided with a complete set of cleaning tools, each securely held in its proper place, and fitted with lock and key, clamps, and conveniently arranged handles.

57. Each case shall contain the following:

One carpet renovator, with slot ¼ in. by 15 in.

One bare floor renovator, 15 in. long, with curved felt-covered face.

One wall brush, with skirted bristles, 12 in. long and ½ in. wide.

One hand brush, with hose connections at end, 8 in. long, 2 in. wide.

One 4-in. round brush for relief work.

One upholstery renovator.

One corner cleaner.

One radiator tool.

One curved stem about 5 ft. long.

One extension tube 5 ft. long.

At least one hat brush with the system.

58 to 64. Same as for Class 1.

64a. All brushes to be of substantial construction, with best quality bristles set in close rows and as thick as possible, skirted with rubber, leather, or chamois skin, so that all air entering renovator will pass over surface being cleaned.

65 to 68. Same as for Class 1.

69. *Hose Racks.* Furnish and properly secure in place, where directed, ... hose racks in basement, ... each in first and second stories (.... racks in all). The racks to be constructed of cast iron, galvanized or enamel finish, and each rack to be suitable for holding 75 ft. of hose of required size.

69a. *Hose.* There must be furnished with each hose rack 75 ft. of noncollapsible hose in three 25-ft. lengths.

70 to 73. Same as for Class 1.

74. On completion of the plant the pump will be operated with all outlets closed, and, under this condition, the power consumption must not be more than 50% of that required under test conditions.

75. *Test of Separators.* At each of——points, near——outlets on different risers selected by the architect's representative, the contractor shall furnish and spread on the floor, evenly covering an area of approximately 50 sq. ft. for each outlet, a mixture of 6 lbs. of dry sharp sand that will pass a 50-mesh screen, 3 lbs. of fine wheat flour and 6 lbs. of Portland cement.

75b. Fifty feet of hose shall be attached to each of the—— outlets, and the surfaces prepared for cleaning shall be cleaned simultaneously by operators provided by the contractor until all of the sand, flour and Portland cement has been taken up, when the exhauster shall be stopped and the dirt removed from the separator and spread on the floor again, and the operation of cleaning repeated until the mixture has been handled by the apparatus four times.

The bag contained in the separator must not be disturbed until after completion of the capacity test, which will be made with material in place in the separator after being picked up the fourth time. After completion of the capacity test the contents of separator will be removed. During test of separators a dampened cloth will be held over the exhaust from pump. If such cloth indicates dirt passing through the separator, same will be rejected.

76. To test the capacity of the plant, one hose line 100 ft. long shall be attached to inlet farthest from the separator with standard vacometer, with ½-in. opening in end of hose. .... hose lines shall be attached to other outlets, each with 50 ft. hose and vacometers in end of hose, ... vacometers having ½-in. opening and .... vacometers having ⅞-in opening. Under these conditions 4 in. mercury must be maintained in vacometer at end of 100 ft. of hose.

77 and 78. Same as for Class 1.

Modifications of Specifications when Alternating Current is Available.—When alternating current is available, instead of direct, modify specifications as follows:

23. Motor to be wound for .... volts, ... cycle, ... phase alternating current.

23a. Bidders must name efficiency and power factor of motor at one-half and full load.

24. Motor to have phase-wound rotor with collector rings for insertion of starting resistance.

Omit 25, 26 and 27.

28. Same as for direct current.

29. There must be an insulation between the starter or primary windings and the frame of not less than one megohm.

30, 31, 32, 32a. Same as for direct current.

33. *Rheostat.* Furnish and install an approved hand-starting rheostat for inserting resistance in rotor circuit in starting, of proper size to insure the starting of motor in not exceeding 15 seconds without overheating.

33a. Same as for direct current, except that switch must be either three- or four-pole, according to current available.

Omit 36d with alternating current machine.

CLASS 3

LARGE INSTALLATION WITH UNUSUALLY LONG PIPE LINES.

1. Same as for Class 1.
2. Exhauster shall be of the reciprocating piston type.
3. The piston type of exhauster shall be double acting and so designed that the cylinder clearance shall be reduced to a minimum, or suitable device shall be employed to minimize the effect of large clearance.
4. The induction and eduction valves may be either poppet, rotary, or semi-rotary, and shall operate smoothly and noiselessly.
5. The piston packing shall be of such character as to be practically air tight under working conditions and constructed so that it will be set out with its own elasticity without the use of springs of any sort. If metallic rings are used, they must fill the grooves in which they are fitted, both in width and depth, and must be concentric; that is, of the same thickness throughout. The joint in the ring or rings to be lapped in width but not in thickness, and if more than one ring is used they are to be placed and doweled in such position in their respective grooves so that the joints will be at least one-fourth of the circumference apart.
6. The piston shall have no chamber or space into which air may leak from either side of the piston. All openings into the body of the piston must be tightly plugged with cast-iron plugs.
7. The piston rod stuffing box to be of such size and depth that if soft packing is used it can be kept tight without undue pressure from the gland. If metallic packing is used, it must be vacuum tight without undue pressure on the rod. Proper means shall be provided for the continuous lubrication of the piston rod.
8. The exhauster of the piston type shall be fitted with an approved cross-head suitably attached to the piston rod; ma-

chines having an extended piston rod for guide purposes will not be acceptable.

Omit 9 to 13.

14. Same as for Class 1.

15. Reciprocating piston exhauster shall be provided with the necessary devices for the removal of the heat generated by friction and compression, that shall prevent the temperature of cylinders or eduction chambers rising more than 100° F. above the surrounding atmosphere after two hours' continuous operation under full-load conditions.

16. *Speed.* Reciprocating exhauster with poppet valves shall operate at an average piston speed not exceeding 200 ft. per minute, with rotary valves not exceeding 300 ft. per minute.

Omit 17 and 18.

18a. *Base Plate, Foundation, etc.* Provide suitable base plate to rigidly support the exhauster and its motor as a unit, which shall be large enough to catch all drip of water or oil. Provide a raised margin and pads for feet of exhauster frame, motor, and anchor bolts, high enough to prevent any drip from getting into the foundation or on the floor.

18b. Provide suitable foundation of brick or concrete, to which base plate shall be firmly anchored. The foundation shall be built on top of the cement floor of the basement, which shall be picked to afford proper bond for the foundation.

18c. Construct the foundation of such a height as to bring the working parts of the machine at the most convenient level for operating purposes. Exposed parts of the foundation to be faced with best grade white enameled brick. If the base plate does not cover the foundation, the exposed top surface is to be finished with enameled brick, using bull-nose brick on all edges and corners.

19 to 23. Same as for Class 1.

23a. The guaranteed efficiency of motor shall not be less than 80% at half load and not less than 85% at full load.

24 to 32. Same as for Class 1.

32a. Motor shall be subject to shop test to determine efficiency, heating, insulation, etc. Manufacturers' certified test sheets of motor, giving all readings taken during shop test, together with calculated results, must be submitted to the Architect for approval before motor is shipped from factory.

33. *Rheostat.* Furnish and install where shown, upon a slate panel hereinafter specified, a starting rheostat of proper size and approved make, designed for the particular duty it has to perform. It must have an automatic no-voltage and overload release. All resistance for rheostat is to be placed on the back of the tablet. Contacts must project through board to front side. All moving parts must be on front of board.

33a. *Tablet.* Furnish and place where shown, a slate tablet, not less than ¾ in. thick, supported by a substantial angle bar frame, so placed that there will be a space of not less than 4 in. between the wall and back of resistance. Mount on this tablet one double-pole, 250-volt knife switch, with two 250-volt inclosed fuses and one starting rheostat, as specified hereinbefore. The connections shall be on the back of the tablet. The space between the column and the tablet shall be inclosed with a removable diamond-mesh grill of No. 10 iron wire in channel frame.

34, 35 and 36. Same as for Class 1.

37. *Dust Separators.* There shall be one dry and one wet separator located where shown on drawings. Each separator shall have a volume of 3 cu. ft. for each renovator of plant capacity.

38. The separator first receiving the dust shall be a dry separator, the interior arrangement of which shall be such that no part of same shall receive the direct impact of the dust. No screens or cloth bags shall be used in this separator and it must be so constructed that it will intercept 95% of the dust entering same.

38a. The second separator must be a wet separator which may be contained in the base of the machine or consist of a separate tank.

38b. Wet separators, whether separate from or integral with the base of the machine, must be provided with an attachment which will positively mix the air and water, thoroughly break up all bubbles, separate the water from the air, and prevent any water entering the exhauster cylinder.

38c. Suitable means must be provided to automatically equalize the vacuum between wet and dry separators upon the shutting down of the exhauster.

38d. The separators must be provided with suitable openings for access to the interior for inspection and cleaning, and the interior arrangement of the separators must be such that they may be readily cleaned without dismantling.

38e. All parts of the wet separator tank not constructed of non-corrosive metal must be thoroughly tinned or galvanized both inside and outside. The interior of the wet separator formed in base of exhauster shall be painted with at least two coats of asphalt varnish or other paint suitable to prevent the corrosion of same.

38f. Separators must be provided with all necessary valves or other attachments for successful operation, including a sight glass for the wet separator, through which the interior of the same may be observed, and an iron-case mercury column reading 50% in excess of operating vacuum, attached to the dry separator first receiving the dust.

38g. The wet separator shall be properly connected to water supply where directed and discharge to sewer where shown on plans.

38h. A running trap with clean-out shall be installed in the waste line.

39 to 41. Same as for Class 1.

41a. Waste and water pipe, in connection with wet separator and jacket, except waste pipe below basement floor, to be standard galvanized wrought-iron pipe or steel screw-jointed pipe free from burs. Waste pipe below the basement floor is to be best grade, "extra heavy" cast-iron pipe, with lead-calked joints.

42 to 45. Same as for Class 1.

45a. Fittings on water lines to be standard galvanized beaded fittings.

45b. Fittings on waste line above basement floor line to be galvanized recessed screw-jointed drainage fittings and those below basement floor to be "extra heavy" cast-iron with hub joints.

46 to 50. Same as for Class 1.

50a. The exhaust pipe is to be fitted with an approved first-class exhaust muffler not less than 12 in. in diameter and 60 in. high, closely riveted and constructed of galvanized iron not less than 1/8 in. thick, and in event an exhauster requiring lubrication is furnished, this muffler is to be arranged so that it will also be an efficient oil separator. Drip connection to be arranged at bottom of muffler.

51 to 56. Same as for Class 1.

56a. *Tool Cases.* Furnish .... approved hardwood cabinet-finished cases for cleaning tools. Each case to be made as light as possible and of convenient form for carrying by hand and provided with a complete set of cleaning tools, each securely held in its proper place, and fitted with lock and key, clamps and conveniently arranged handles.

57. Each case shall contain the following:

One carpet renovator, with slot 1/4 in. by 12 in.

One bare floor renovator 12 in. long, with curved, felt-covered face.

One wall brush, with skirted bristles, 12 in. long and 1/2 in. wide.

One hand brush, with hose connection at end, 8 in. long and 2 in. wide.

One 4-in. round brush for relief work.

One upholstery renovator.

One corner cleaner.

One radiator tool.

One curved stem about 5 ft long.

One straight extension stem 5 ft. long.

At least one hat brush with the system.

58 to 64. Same as for Class 1.

64a. All brushes to be of substantial construction, with best quality bristles set in close rows and as thick as possible, skirted with rubber, leather, or chamois skin, so that all air entering renovator will pass over surface being cleaned.

65 to 68. Same as for Class 1.

69. *Hose Racks.* Furnish and properly secure in place where directed, ... hose racks in basement, ... each in first and

second stories (.... racks in all). The racks to be constructed of cast-iron, galvanized or enamel finish, and each rack to be suitable for holding 75 ft. of hose of required size.

69a. *Hose.* There must be furnished with each hose rack 75 ft. of non-collapsible hose in three 25-ft. lengths.

70. Hose to be 1 in. inside diameter of best quality, rubber hose, reinforced in best manner to absolutely prevent collapse at highest vacuum obtainable with the exhauster furnished and to prevent collapse if stepped on. Weight of hose to be not over 12 oz. per linear foot.

71, 72 and 73. Same as for Class 1.

74. On completion of the plant the pump will be operated with all outlets closed and, under this condition, the power consumption must not be more than 50% of that required under test conditions.

75. To test the capacity of the plant, ... hose lines each 100 ft. long will be attached to outlets on the system and each hose fitted with a standard vacometer. .... vacometers shall have ½-in. opening and .... vacometers shall have ⅞-in. opening. Under these conditions 4 in. vacuum must be maintained at vacometers having ½ in. opening.

75a. *Test of Separators.* At each of——points, near——outlets on different risers selected by the architect's representative, the contractor shall furnish and spread on the floor, evenly covering an area of approximately 50 sq. ft. for each outlet, a mixture of 6 lbs. of dry sharp sand that will pass a 50-mesh screen, 3 lbs. of fine wheat flour, and 1 lb. of finely pulverized charcoal.

75b. Fifty feet of hose of size required by the system shall be attached to each of the——outlets, and the surface or surfaces prepared for cleaning shall be cleaned simultaneously by operators provided by the contractor until all of the sand, flour and charcoal has been taken up, when the exhauster shall be stopped and the dirt removed from the dry separator and spread on the floor again, and the operation of cleaning repeated until the mixture has been handled by the apparatus four times. If, after thoroughly flushing the system at completion of above run, any dust or mud is found in the cylinder, ports, or valve chambers of the displacement exhauster, or if less than 95% of the dirt removed is found in the dry separator, it shall be deemed sufficient ground for the rejection of the separators.

76 and 77. Same as for Class 1.

Modifications of Specifications when Alternating Current is Available.—When alternating current is available, instead of direct, modify specifications as follows:

23. Motor to be wound for .... volts, ... cycle, ... phase alternating current.

23a. Bidders must name efficiency and power factor of motor at one-half and full load.

24. Motor to have phase-wound rotor with collector rings for insertion of starter resistance.

Omit 25, 26 and 27.

29. Same as for Class 1, alternating current.

33. *Rheostat.* Furnish and install an approved hand-starting rheostat for inserting resistance in rotor circuit in starting, of proper size to insure the starting of motor in not exceeding 15 seconds without overheating.

33a. Same as for direct current, except switch must be either three- or four-pole, according to current available.

CLASS 4

LARGE OR SMALL PLANT WHERE CARPET CLEANING IS OF SECONDARY IMPORTANCE.

1 to 17. Same as for Class 1.

Omit 18.

18a. *Base Plate, Foundation, etc.* Provide suitable base plate to rigidly support the exhauster and its motor as a unit, which shall be large enough to catch all drip of water or oil. Provide a raised margin and pads for feet of exhauster frame, motor, and anchor bolts, high enough to prevent any drip from getting into the foundation or on the floor.

18b. Provide suitable foundation of brick or concrete, to which the base plate shall be firmly anchored. The foundation shall be built on top of the cement floor of the basement, which shall be picked to afford proper bond for the foundation.

18c. Construct the foundation of such a height as to bring

the working parts of the machine at the most convenient level for operating purposes. Exposed parts of the foundation to be faced with best grade white enameled brick. If the base plate does not cover the foundation, the exposed top surface is to be finished with enameled brick, using bull-nose brick on all edges and corners.

19 to 23. Same as for Class 1.

23a. The guaranteed efficiency of motor shall not be less than 78% at half load and not less than 84% at full load.

24 to 32. Same as for Class 1.

32a. Motor shall be subject to shop test to determine efficiency, heating, insulation, etc. Manufacturers' certified test sheets of motor, giving all readings taken during shop test, together with calculated results, must be submitted to the Architect for approval before motor is shipped from factory.

33. *Rheostat.* Furnish and install where shown, upon a slate panel hereinafter specified, a starting rheostat of proper size and approved make, designed for the particular duty it has to perform. It must have an automatic no-voltage and overload release. All resistance for rheostat is to be placed on the back of the tablet. Contacts must project through board to front side. All moving parts must be on front of board.

33a. *Tablet.* Furnis hand place where shown, a slate tablet not less than ¾ in. thick, supported by a substantial angle iron frame, so placed that there will be a space of not less than 4 in. between the wall and back of resistance. Mount on this tablet one double-pole, 250-volt knife switch, with two 250-volt inclosed fuses and one starting rheostat, as specified hereinbefore. The connections shall be on the back of the tablet. The space between the column and the tablet shall be enclosed with a removable diamond-mesh grill of No. 10 wire in channel frame.

34, 35 and 36. Same as for Class 1.

36a. *Automatic Control.* Suitable means shall be provided in connection with the rotary exhauster that will maintain the vacuum in the separators within the limit of the machine at point found to be most desirable, irrespective of the number of sweepers in operation.

36b. Controller shall consist of a suitable means provided in the exhauster, or as an attachment thereto, which will automatically throw the exhauster out of action by admitting atmospheric pressure to the exhauster only, but not to the system; whenever the vacuum in the separator rises above the point considered desirable, and throw the exhauster into action when the vacuum falls below the established lower limit.

36c. In addition to control, a positive vacuum breaker having an opening equal to 1 in. diameter pipe net for 6 in. of mercury, must be provided on separator.

36d. If centrifugal fan is used, no control or vacuum breaker will be required.

37. Furnish one separator having a cubic contents of 4.5 cu. ft. for each sweeper of plant capacity.

38 to 56. Same as for Class 1.

56a. *Tool Cases.* Furnish approved hardwood cabinet-finished cases for cleaning tools. Each case to be made as light as possible and of convenient form for carrying by hand and provided with a complete set of cleaning tools, each securely held in its proper place, and fitted with lock and key, clamps, and conveniently arranged handles.

57. Each case shall contain the following:

One carpet renovator ½ in. by 15 in.

One bare floor renovator, 15 in. long, with curved felt-covered face.

One wall brush, with skirted bristles, 12 in. long and ½ in. wide.

One hand brush, with hose connection at end, 8 in. long and 2 in. wide.

One 4-in. round brush for relief work.

One upholstery renovator.

One corner cleaner.

One radiator tool.

One curved stem about 5 ft. long.

One straight extension stem 5 ft. long.

At least one hat brush with the system.

58 to 68. Same as for Class 1.

69. *Hose Racks.* Furnish and properly secure in place, where

directed, ... hose racks in basement, ... each in first and second stories (.... racks in all). The racks to be constructed of cast-iron, galvanized or enamel finish, and each rack to be suitable for holding 75 ft. of hose of required size.

69a. *Hose.* There must be furnished with each hose rack 75 ft. of non-collapsible hose in three 25-ft. lengths.

70. Hose to be 1½ in. or 1¾ in. inside diameter, best quality rubber hose, reinforced in best manner to absolutely prevent collapse at highest vacuum obtainable with the exhauster furnished and to prevent collapse if stepped on. Weight of hose to be not over 12 oz. per linear foot.

71 to 73. Same as for Class 1.

74. On completion of the plant the pump will be operated with all outlets closed and, under this condition, the power consumption must not be more than 50% of that required under test conditions.

75. Same as for Class 1.

76. To test the capacity of the plant, ... hose lines each 75 ft. long shall be attached to the inlets, each hose to be fitted with standard vacometer with ⅞-in. opening. Under these conditions a vacuum of 1 in. mercury must be maintained in each vacometer.

77 and 78. Same as for Class 1.

CLASS 5

To Give Widest Competition.

1. Same as for Class 1.

2. Exhauster to be piston, rotary or centrifugal fan type.

3. The piston type of exhauster shall be double-acting and so designed that the cylinder clearance shall be reduced to a minimum, or suitable devices shall be employed to minimize the effect of large clearances.

4. The induction and eduction valves may be either poppet, rotary or semi-rotary, and shall operate smoothly and noiselessly.

5. The piston packing shall be of such character as to be practically air tight under working conditions and constructed so that it will be set out with its own elasticity without the use

of springs of any sort. If metallic rings are used, they must fill the grooves in which they are fitted, both in width and depth, and must be concentric; that is, of the same thickness throughout. The joint in the ring or rings to be lapped in width but not in thickness, and if more than one ring is used they are to be placed and doweled in such position in their respective grooves so that the joints will be at least one-fourth of the circumference apart.

6. The piston shall have no chamber or space into which air may leak from either side of the piston. All openings into the body of the piston must be tightly plugged with cast-iron plugs.

7. The piston-rod stuffing box to be of such size and depth that if soft packing is used it can be kept tight without undue pressure from the gland. If metallic packing is used, it must be vacuum tight without undue pressure on the rod. Proper means shall be provided for the continuous lubrication of the piston rod.

8. The exhauster of the piston type shall be fitted with an approved cross-head suitably attached to the piston rod; machines having an extended piston rod for guide purposes will not be acceptable.

Insert paragraphs 3 to 15 from specifications for Class 1.

15a. Reciprocating exhauster shall be provided with the necessary devices for the removal of the heat generated by friction and compression, that shall prevent the temperature of cylinders or eduction chambers rising more than 100° F. above the surrounding atmosphere after two hours' continuous operation under full load conditions.

15b. *Speed.* Reciprocating exhauster with poppet valves shall operate at an average piston speed not exceeding 200 ft. per minute, with rotary valves not exceeding 300 ft. per minute.

Insert paragraphs 16 and 17 from specifications for Class 1. Omit 18.

18a. *Base Plate, Foundation, etc.* Provide suitable base plate to rigidly support the exhauster and its motor as a unit, which shall be large enough to catch all drip of water or oil. Provide a raised margin and pads for feet of exhauster frame, motor, and anchor bolts, high enough to prevent any drip from getting into the foundation or on the floor.

18b. Provide suitable foundation of brick or concrete, to which the base plate shall be firmly anchored. The foundation shall be built on top of the cement floor of the basement, which shall be picked to afford proper bond for the foundation.

18c. Construct the foundation of such a height as to bring the working parts of the machine at the most convenient level for operating purposes. Exposed parts of the foundation to be faced with best grade white enameled brick. If the base plate does not cover the foundation, the exposed top surface is to be finished with enameled brick using bull-nose brick on all edges and corners.

19 to 23. Same as for Class 1.

23a. The guaranteed efficiency of motor shall not be less than 78% at half load and not less than 84% at full load.

24 to 32. Same as for Class 1.

32a. Motors shall be subject to shop test to determine efficiency, heating, insulation, etc. Manufacturers' certified test sheets of motor, giving all readings taken during shop test, together with calculated results, must be submitted to the architeet for approval before motor is shipped from factory.

33. *Rheostat.* Furnish and install where shown, upon a slate panel hereinafter specified, a starting rheostat of proper size and approved make, designed for the particular duty it has to perform. It must have an automatic no-voltage and overload release. All resistance for rheostats is to be placed on the back of the tablet. Contacts must project through board to front side. All moving parts must be on front of board.

33a. *Tablet.* Furnish and place where shown, a slate tablet not less than ¾ in .thick, supported by a substantial angle bar frame, so placed that there will be a space of not less than 4 in. between the wall and back of resistance. Mount on the tablet one double-pole, 250-volt knife switch, with two 250-volt enclosed fuses and one starting rheostat, as specified hereinbefore. The connections shall be on the back of the tablet. The space between the column and the tablet shall be enclosed with a removable diamond-mesh grill of No. 10 iron wire in channel frame.

34, 35 and 36. Same as for Class 1.

36a. *Automatic Control.* Suitable means shall be provided in connection with the reciprocating and rotary exhausters that will maintain the vacuum in the separators within the limit of the machine at point found to be most desirable, irrespective of the number of sweepers in operation.

36b. Controller shall consist of a suitable means provided in the exhauster, or as an attachment thereto, which will automatically throw the exhauster out of action by admitting atmospheric pressure to the exhauster only, but not to the system; or that shall cause suction from the system to cease whenever the vacuum in the separators rises above the point considered desirable, and throw the exhauster into action when the vacuum falls below the established lower limit.

36c. *Vacuum Breaker.* In addition to the controlling devices above specified, if a reciprocating or rotary exhauster is used, there shall be placed in the suction pipe to the exhauster an approved positive-acting vacuum breaker having opening equivalent to the area of 1-in. diameter pipe and set to open at 12 in.

36d. If centrifugal fan is used, no control or vacuum breaker will be required.

37. *Dust Separators.* There must be provided at least one separator between the pipe lines and exhauster having a volume of not less than 3 cu. ft. per sweeper of plant capacity. This separator must be so constructed that no part thereof will receive the direct impact of the dust. If rotary exhauster is used, this separator must also contain a bag so placed that only the lightest dust will reach same and must be arranged to be readily cleaned without dismantling the separator. If a centrifugal exhauster is used, this apparatus may or may not contain a bag, and, if piston pump is used, this separator must contain no bags or screens whatever. If a piston type of exhauster is installed, an additional separator must be placed between the first separator and the exhauster. This must be a wet separator and may be contained in the base of the machine or consist of a separate tank.

Omit 38.

38a. Same as for Class 1.

38b. Wet separators, whether separate from or integral with the base of the machine, must be provided with an attachment which will positively mix the air and water, thoroughly break up all bubbles, separate the water from the air, and prevent any water entering the exhauster cylinder.

38c. Suitable means must be provided to automatically equalize the vacuum between wet and dry separators upon the shutting down of the exhauster.

38d. The separators must be provided with suitable openings for access to the interior for inspection and cleaning, and the interior arrangement of the separators must be such that they may be readily cleaned without dismantling.

38e. All parts of the wet separator tank (if used) not constructed of non-corrosive metal must be thoroughly tinned or galvanized both inside and outside. The interior of the wet separator formed in base of exhauster shall be painted with at least two coats of asphalt varnish or other paint suitable to prevent the corrosion of same.

38f. Separators must be provided with all necessary valves or other attachments for successful operation, including a sight glass for the wet separator (if used), through which the interior of same may be observed.

38g. The wet separator (if used) shall be properly connected to water supply where directed and discharge to sewer where shown on plans.

38h. A running trap with clean-out shall be installed in the waste line.

39 to 41. Same as for Class 1.

41a. Waste and water pipe, in connection with wet separator and jacket, except waste pipe below basement floor, to be standard galvanized wrought-iron or steel screw-jointed pipe free from burs. Waste pipe below the basement floor is to be best grade, "extra heavy" cast-iron pipe, with lead-calked joints.

42 to 45. Same as for Class 1.

45a Fittings on water lines to be standard galvanized beaded fittings.

45b. Fittings on waste line above basement floor to be galvanized recessed screw-jointed drainage fittings and those below basement floor to be "extra heavy" cast-iron with hub joints.

46 to 50. Same as for Class 1.

50a. If reciprocating exhauster is used, the exhaust pipe is to be fitted with an approved first-class muffler not less than 12 in. in diameter and 60 in. high, closely riveted and constructed of galvanized iron not less than 1/8 in. thick, and in event an exhauster requiring lubrication is furnished this muffler is to be arranged so that it will also be an efficient oil separator. Drip connection is to be arranged at bottom of muffler.

51 to 56. Same as for Class 1.

56a. *Tool Cases.* Furnish .... approved hardwood cabinet-finished cases for cleaning tools. Each case to be made as light as possible and of convenient form for carrying by hand and provided with a complete set of cleaning tools, each securely held in its proper place, and fitted with lock and key, clamps, and conveniently arranged handles.

57. Each case will contain the following:

One carpet renovator, with slot 1/4 in. by not less than 12 or more than 15 in. long.

One bare floor renovator, 15 in. long, with curved felt-covered face.

One wall brush, with skirted bristles, 12 in. long and 1/2 in. wide.

One hand brush, with hose connection at end, 8 in. long and 2 in. wide.

One 4-in. round brush for relief work.

One upholstery renovator.

One corner cleaner.

One radiator tool.

One curved stem about 5 ft. long.

One straight extension stem 5 ft. long.

At least one hat brush with the system.

58 to 64. Same as for Class 1.

64a. All brushes to be of substantial construction, with best quality bristles set in close rows and as thick as possible, skirted with rubber, leather, or chamois skin, so that all air entering renovator will pass over surface being cleaned.

65 to 68. Same as for Class 1.

69. *Hose Racks.* Furnish and properly secure in place, where directed, ... hose racks in basement, ... each in first and second stories (.... racks in all). The racks to be constructed of cast-iron, galvanized or enamel finish, and each rack to be suitable for holding 75 ft. of hose of required size.

69a. *Hose.* There must be furnished with each hose rack, 75 ft. of non-collapsible hose in three 25-ft. lengths.

70. Hose shall not be less than 1 in. or more than 1¾ in. inside diameter, best quality rubber hose, reinforced in best manner to absolutely prevent collapse at highest vacuum obtainable with the exhauster furnished and to prevent collapse if stepped on. Weight of hose to be not over 12 oz. per linear foot.

71 to 73. Same as for Class 1.

74. On completion of the plant, the pump will be operated with all outlets closed and under this condition the power consumption must not be more than 50% of that required under test conditions.

75. To test the capacity of plant, one hose line 100 ft. long shall be attached to inlet farthest from the separator, with standard vacometer with ½-in. opening in end of hose. Hose lines shall be attached to other outlets, each with 50-ft. hose and vacometer in end of hose, ... vacometers having ½-in. opening and .... vacometers having ⅞-in. opening. Under these conditions 4 in. mercury must be maintained in vacometer at end of 100 ft. of hose.

75a. *Test of Separators.* At each of——points, near—— outlets on different risers selected by the ...... representative, the contractor shall furnish and spread on the floor, evenly covering an area of approximately 50 sq. ft. for each outlet, a mixture of 6 lbs. of dry sharp sand that will pass a 50-mesh screen, 3 lbs. of fine wheat flour, and 1 lb. of finely pulverized charcoal, if wet separator be used, and 6 lbs. of Portland cement, if bag be used.

75b. Fifty feet of hose of size required by the system used

shall be attached to each of the——outlets, and the surface or surfaces prepared for cleaning shall be cleaned simultaneously by operators provided by the contractor until all of the sand, flour and charcoal has been taken up, when the exhauster shall be stopped and the dirt removed from the dry separator and spread on the floor again, and the operation of cleaning repeated until the mixture has been handled by the apparatus four times. If, after thoroughly flushing the system at completion of the above run, any dust or mud is found in the cylinder, ports, or valve chambers of the displacement exhauster, or if less than 95% of the dirt removed is found in the dry separator of the centrifugal exhauster, it shall be deemed sufficient ground for the rejection of the separators. If bag is used, same must not be disturbed until after capacity test, which will be made with material in separator after being picked up the fourth time.

76-77. Same as for Class 1.

78. Evaluation of proposal (for 4-sweeper plant): No proposal will be considered which contemplates furnishing an exhauster requiring more power to operate under test conditions than:

Full load, 14 K. W.; three-quarter load, 12.25 K. W.; one-half load, 10.5 K. W.

Test Requirement: Paragraph 75 to be considered full load. To reduce the load to three-quarters, one ⅞-in. vacometer opening to be closed; to produce one-half load, one ½-in. opening to be closed in addition to the ⅞-in. opening.

Bidders are requested to state in proposal the power consumption required by their apparatus at full, three-quarters and one-half loads, and in case the guarantees of the various bidders differ, they will be evaluated as follows for the purpose of comparison:

For each full K. W. of power consumption or fractional part thereof there will be allowed the following amount or proportionate parts for fractional parts of a K. W. hour at the various loads:

Per Cent. of Full Load.	100	75	50
Amount	$156.00	$62.00	$94.00

As an illustration, let it be assumed that proposals have been received offering equipment in accordance with specification requirements, and the one offering the most economical apparatus based on guaranteed power consumption names the highest price for the installation.

To determine if the purchaser will be justified in accepting the highest proposal, let it be assumed that he has guaranteed a power consumption of 1 K. W. less at full load, 0.75 K. W. less at three-quarters load, and 0.25 K. W. more at half-load than the lowest bidder. Under these conditions the algebraic sum of the saving of the higher bidder over the lower bidder would be 156+46.50—23.50=179, which is the additional amount in dollars which the purchaser would be warranted in paying for the apparatus of higher efficiency.

In making the economy test to determine if the guaranteed power consumption has been fulfilled, an integrating watt meter, previously calibrated and found correct, will be placed in circuit and a two-hour run made at each load and the power consumption based on the meter readings.

Penalty.—It must be distinctly understood to be one of the conditions under which bids are to be submitted for the work embraced in this specification that the apparatus shall meet every requirement of the specification and the guaranty for efficiency under which conditions the contract price will be paid. In the event the apparatus tested fails to meet the specified requirements for capacity or economy, or both, the Architect shall have the right to reject the apparatus, absolutely, and require the installation of satisfactory apparatus, which shall comply with the contract requirements; or if he elects to accept the same, in the event the capacity or efficiency at any load (irrespective of other loads) is less than that named in the proposal, then the contract price shall be the amount named in the contract for a satisfactory plant, less the amount of deficiencies shown by the test based on the following:

For Capacity.—$500.00 for each inch of vacuum and a proportionate part thereof for each fraction of an inch below the 4 in. required in vacometer when operating under full load.

For Economy.—Deduction for each K. W. or a proportionate part thereof for each fraction thereof required in excess of guarantee.

Percentage of Full Load	100	75	50
Penalty	$229.00	$93.00	$141.00

This evaluation was based on the same time of operation and cost of current as that used in illustration under tests (Chapter XII).

The maximum power to be allowed for plants of various capacities should be as follows:

Capacity in Sweepers	100% of Load	75% of Load	50% of Load
8	24	20	17
6	20	18	15
4	14	12.25	10.5
2	7.5	.....	6.25

In event that the plant is to be run with vacuum "on tap," as in a hotel, a guaranteed power consumption at no load should be required and evaluated on the number of hours the plant will probably operate under these conditions. This will be the largest item in the evaluation under such conditions.

CHAPTER XV.

PORTABLE VACUUM CLEANERS.

While this book is primarily intended to deal only with vacuum cleaning systems, which would limit the work to such apparatus as is permanently installed within the building to be cleaned, the author considers that it would not be complete without some mention of the portable cleaners which are so popular at the present time.

On first consideration, the portable cleaner would appear to have a considerable advantage over the stationary type in that the length of hose is usually limited to not over 15 ft. and there is no pipe line, which results in the elimination of practically all friction loss, giving practically the same vacuum at the renovator as at the exhauster. This should result in a saving of practically 50% of the power required to operate the exhauster.

Referring to Chapter XII, we find that the power required to operate a really efficient vacuum cleaning system is approximately 2.5 H. P. per sweeper. If a portable cleaner, with the same efficiency and capacity, be built, it would require at least 1½ H. P.

Such a cleaner would not be portable in the sense of the term as applied to the most popular cleaners today. The same type has been built on special order by the American Radiator Company, which mounted its 1½ H. P. Arco Wand machine on a truck. This cleaner weighs several hundred pounds and could be moved up and down stairs about as easily as a sewing machine and would not be of any service in a building not equipped with elevators. The power required to operate this cleaner is also so great that special power wiring and large capacity outlet plugs have to be installed throughout the building. Such equipment has been provided in at least two department stores where these cleaners are in use. This means that one wires his building for vacuum cleaning instead of piping it, and there is also the necessity of moving a heavy machine about to do the same work as a stationary plant.

It would appear to the author that the cost of wiring would about equal that of piping and that the additional labor required to move the machine about would cost as much as the additional power needed by the stationary exhauster.

This cleaner, as well as all other portable cleaners, discharges the air from the exhauster directly back into the apartment cleaned, and is open to the same objection that was raised against the early compressed air cleaners. While all the dust may be caught by the dust bag, the microbes are allowed to escape with the air and the cleaner is not a sanitary device by any manner of means.

There are a few portable machines using rotary exhausters of the Root type, and piston pumps, all of which are heavy to move about and, in making them as light as possible, the efficiency of the exhauster has been sacrificed. These machines will do the same quality of cleaning as the stationary plants recommended for residence work and they require about ¾ H. P., which is no less than is needed for a stationary plant of the same capacity and efficiency.

The most popular type of portable cleaner is one which can be attached to a socket or plug connected with the lighting system. This should limit the power consumption to ⅛ H. P. However, many of these cleaners use as much as 400 watts and a fair average for cleaners retailing at about $125.00 is 250 watts. Such cleaners will exhaust about 25 cu. ft. of air with a vacuum of 1 in. mercury at the vacometer, a ⅝-in. orifice being used. The theoretical power required to move the air is approximately 50 watts and the overall efficiency of these cleaners is, therefore, about 20%, as against 40% to 50% in a good, one-sweeper stationary plant. The power expended in operating these portable cleaners in proportion to the work done is no less than with an efficient stationary plant.

Portable cleaners have been made in many types but practically all the standard makes use one or two forms of vacuum producers, either the diaphragm pump or the single or multi-stage fan. The pumps of the former type are able to produce a vacuum as high as 6 in. to 10 in. of mercury, when no air is passing, and will displace as high as 30 cu. ft. of free air per minute, when operated with a free inlet. They produce about 1 in. of mercury at the carpet renovator when operated on an ordinary carpet. When small-sized upholstery renovators are used, a much higher vacuum is possible. When operated with bare floor renovators or brushes, the quantity of air exhausted is not much over 20 cu. ft. per minute and they make very inefficient bare floor and wall cleaners, but will do thorough carpet and upholstery cleaning provided a small enough renovator is used.

Machines using a multi-stage fan will produce a maximum vacuum of approximately 2 in. of mercury when exhausting no air, and will produce a vacuum of approximately 1 in. of mercury when operated on an ordinary carpet. With an unrestricted inlet, they will exhaust from 40 to 50 cu. ft. of air per minute. When operated on a bare floor, they will exhaust approximately 30 cu. ft. of free air per minute. They are, therefore, more efficient floor cleaners than the pumps, but cannot do thorough carpet and upholstery cleaning, no matter how small the renovator.

The smaller-fan type of machines, in which the fan is placed integral with the carpet renovator and in which hose is not used in cleaning floors or carpets, are provided with a single-stage fan. They produce a suction of not exceeding ½ in. of mercury when no air is exhausted and will exhaust from 5 to 10 cu. ft. of free air per minute when operated on a carpet. With a free inlet they will exhaust from 15 to 20 cu. ft. of free air per minute. These machines are little if any better than ordinary carpet sweepers.

Machines of this type are open to another objection in that the dust bag is placed on the outlet of the fan and the dust in the bag is continually agitated by the passage of the air, with the result that all the finer particles of the dust are blown through the bag back into the apartment. To be effective, the dust bag must always be placed on the suction side of the exhauster and should be so arranged that the dust will not quickly cover the entire area of the bag, for, when this occurs, the suction is quickly reduced to such an extent that no further cleaning can be done until the bag has been cleaned.

There is another type of mechanical cleaner manufactured by the Hoover Suction Sweeper Company which is provided with a mechanically-operated brush for loosening the dirt from the carpet. The dust is then conveyed through a single-stage fan to a dust bag. The cleaner does not depend on the vacuum to loosen the dirt and will do quite effective carpet cleaning with a small expenditure of power. Owing to the small suction produced, it is of little value for cleaning anything but carpets.

From the experience the author has had with portable vacuum cleaners, some thirty makes having been tested for the Treasury Department by him and by the Bureau of Standards, the use of such cleaners is not considered as either an efficient or sanitary means of mechanical cleaning.

If a cleaner requiring small power is required, one of the smaller stationary plants, costing not over $300.00 and operating with ½ or ¾ H. P., is considered a better investment than $125.00 paid for a portable cleaner.

If the purchaser feels that he cannot afford to pay more than $125.00 for his vacuum cleaner, a type such as the Water Witch can be furnished for this price. This cleaner is placed in the basement, with arrangements for starting same from any floor. The manufacturers state that this apparatus produces a vacuum of 2 in. mercury in a carpet renovator, 4 in. mercury in an upholstery renovator and exhausts 25 to 30 cu. ft. of free air per minute with open hose. The machine operates by water pressure and the manufacturers state that it requires about 6 to 8 gals. of water per minute. All air is exhausted outside of the building and all dust washed down the sewer with the exhaust water. It is therefore, a fairly efficient and sanitary cleaning system.

The statements made above apply to parties who own their residences and occupy offices in modern buildings. There are, besides these, a great many who live in rented houses and apartments or occupy offices in buildings where the owners are not sufficiently progressive to install stationary cleaning plants.

To supply the needs of this class is evidently the field of the portable cleaner, as even the poorest of these machines is more effective in the removal of dust and dirt than the broom and carpet sweeper.

The selection of a portable cleaner by one who must necessarily resort to the use of such a cleaner should be made with care. The motor should be looked into and only one which has brushes readily removable and one in which the condition of the brushes can be easily noted should be selected. Lubrication is important. A good cleaner should be so constructed that it can be operated for at least 100 hours without relubrication.

The dust bag should always be on the suction side of the vacuum producer and of such a design and construction that at least ½ peck of a mixture of 40% sand, 30% flour, 15% sweepings and 15% Portland cement can be picked up from the floor and retained in the bag and the machine still be capable of picking up material from a bare floor.

A good test for capacity of a portable machine is to pick up ½ peck of such material, then fit a thin disk with ⅞-in. diameter opening over the end of the hose. A machine, to be of any value, should show a suction of 3 in. water and a first-class machine will show 8 in. under these conditions. This will do fairly good bare floor work. To ascertain if the machine will clean carpets, use a similar disk with ⅝-in. diameter opening, when a suction of 7 in. water indicates the lowest value and 16 in. about the best that can be obtained from any portable cleaner. Cleaners must be readily portable and should not weigh exceeding 75 lbs.

RETURN TO the circulation desk of any
University of California Library

or to the

NORTHERN REGIONAL LIBRARY FACILITY
Bldg. 400, Richmond Field Station
University of California
Richmond, CA 94804-4698

ALL BOOKS MAY BE RECALLED AFTER 7 DAYS

- 2-month loans may be renewed by calling (510) 642-6753
- 1-year loans may be recharged by bringing books to NRLF
- Renewals and recharges may be made 4 days prior to due date

DUE AS STAMPED BELOW

DEC 30 2004

NOV 08 2005

NOV 12 2005

DD20 6M 9-03

FORM NO. DD6

UNIVERSITY OF CALIFORNIA, BERKELEY
BERKELEY, CA 94720

CPSIA information can be obtained
at www.ICGtesting.com
Printed in the USA
BVHW041109300119
539044BV00013B/193/P